AN INTRODUCTION TO FOSSILS AND MINERALS

AN INTRODUCTION TO FOSSILS AND MINERALS

SEEKING CLUES TO THE EARTH'S PAST
REVISED EDITION

JON ERICKSON
FOREWORD BY DONALD R. COATES, PH.D.

☑® Facts On File, Inc.

AN INTRODUCTION TO FOSSILS AND MINERALS
Seeking Clues to the Earth's Past, Revised Edition

Checkmark Books
An imprint of Facts On File, Inc.
132 West 31st Street
New York NY 10001

Library of Congress Cataloging-in-Publication Data

Erickson, Jon, 1948–
    An introduction to fossils and minerals: seeking clues to the earth's
past.—Rev. ed./by Jon Erickson; foreword by Donald Coates.
        p.   cm.—(The living earth)
    Includes bibliographical references and index.
    ISBN 0-8160-4236-5 (hc)   ISBN 0-8160-4237-3 (pb)
    1. Fossils. 2. Minerals. 3. Geology. I. Title.

QE711.2.E75   2000
560–dc21   00-037203

Checkmark Books are available at special discounts when purchased in bulk quantities for businesses, associations, institutions or sales promotions. Please contact our Special Sales Department in New York at 212/967-8800 or 800/322-8755.

You can find Facts On File on the World Wide Web at **http://www.factsonfile.com**

Text design by Cathy Rincon
Cover design by Nora Wertz
Illustrations by Jeremy Eagle and Dale Dyer, © Facts On File

Printed in the United States of America

MP Hermitage 10 9 8 7 6 5 4 3 2
        (PB) 10 9 8 7 6 5 4 3 2 1

This book is printed on acid-free paper.

CONTENTS

TABLES

ACKNOWLEDGMENTS

The author thanks the National Aeronautics and Space Administration (NASA), the National Museums of Canada, the National Oceanic and Atmospheric Administration (NOAA), the National Park Service, the USDA–Soil Conservation Service, the U.S. Geological Survey (USGS), and the Woods Hole Oceanographic Institution (WHOI) for providing photographs for this book.

Special thanks also go to Mr. Frank Darmstadt, Senior Editor at Facts On File, and Ms. Cynthia Yazbek, Associate Editor, for their contributions in the creation of this book.

FOREWORD

INTRODUCTION TO FOSSILS AND MINERALS, REVISED EDITION

No other aspect of geology is as popular with the public as the occurrence of fossils and minerals in earth materials. Countless geologists have been attracted to this discipline because at an early age they discovered the wonder and beauty of such treasures as dinosaurs and precious gems. Jon Erickson has woven a revealing story of the uniqueness of fossils and minerals in the natural world. Indeed this attractive book provides an important source of information for those people who are interested about the world in which they live. This introduction to fossils and minerals admirably fulfills such an objective.

When the reader goes beyond the somewhat bewildering names of fossils, there can be ample reward discovering their importance in helping decipher the mysteries of rock formations and their development throughout time. In a somewhat different way knowledge about minerals also opens a window to their importance. These relationships, and the manner in which they have interacted throughout geological history, are explored in the book's 10 chapters. The author traces Earth's development during the millennia of geologic time (more than 4 billion years) and shows how a knowledge of fossils and minerals is crucial for unraveling this saga.

This short book title does not adequately indicate the large diversity of subject matter that is discussed. Indeed each chapter contains a wealth of topical information that places in perspective the essence of numerous different types of material and their relationships. This systematic treatment of Earth history, rock types, marine fossils, terrestrial fossils, crystals, gems, and precious metals provides indelible insights into Earth's unique character. The data are up-to-date regarding the most recent discoveries on such topics as global tectonics and faunal extinctions, including a wealth of information about the demise of dinosaurs.

The writing style of Jon Erickson is very clear, readable, and understandable to the nonscientist. The accuracy of the book and the scope of information will also be welcomed by geologists. The 178 figures provide visual enhancement, and such data supplement the text descriptions. These include maps, photographs, line drawings, and diagrams. Because the vocabulary of geology may be new to many, a Glossary of word definitions and explanations provides an extra bonus for comprehension. For those who are especially interested in following the documentation of ideas and facts, the Bibliography provides a fine summary. Thus, *An Introduction to Fossils and Minerals* is recommended reading for both those just beginning their discovery journey of the Earth and for scientists who appreciate a well-crafted presentation on these significant subjects.

—Donald R. Coates, Ph.D.

INTRODUCTION

The study of fossils is essential for understanding the mysteries of life, for delineating the evolution of species, and for reconstructing the history of the Earth. The science of geology grew out of the study of fossils, which were used to date various strata. Today, advanced dating techniques enable paleontologists to piece together an accurate picture of the evolution of marine and land animals.

Interest in minerals and gems is evident in the ancient world. Crystals continue to fascinate us with their symmetrical beauty and we depend on the Earth's resources for much of our energy. The rock formations in which minerals are found, reveal, layer by layer, the continual formation and erosion of the Earth's surface. Rocks are also known to whistle in the air, follow the sun's path across the sky, glow in the dark, and reverse their magnetic fields.

This revised and updated edition is a much expanded look at the popular science of paleontology and mineralogy. Readers will enjoy this clear and easily readable text, which is well illustrated with dramatic photographs, clearly drawn illustrations, and helpful tables. The comprehensive Glossary is provided to define difficult terms, and the Bibliography lists references for further reading.

The book is meant to introduce the fascinating science of geology and the way it reveals the history of the Earth as told by its rocks. The text describes the components of the Earth, the different rock types, and the methods with which the fossil and mineral contents of the rocks are found, dated, and classified. It is also designed to aid in the location, identification, and collection of a variety of rock types, many of which contain collectible fossils and minerals. Students of geology and science will find this a valuable reference to further their studies.

Geologic formations can be found in most parts of the United States, often within a short distance from home. On the basis of the information presented here, amateur geologists and collectors will have a better understanding of the forces of nature and the geologic concepts that will help them locate rocks and minerals in the field.

AN INTRODUCTION TO FOSSILS AND MINERALS

1

THE EARTH'S HISTORY
UNDERSTANDING OUR PLANET'S PAST

The Earth is a dynamic planet that is constantly changing. Continents move about on a sea of molten rock. Jagged mountains rise to great heights only to be eroded down to flat plains. Seas fill up and dry out when waters are forced from the land. Glaciers expand across the land and retreat back to the poles. And species evolve and go extinct during times of environmental chaos and change. These episodes in the Earth's history are divided into chunks of time, known as the geologic time scale. Each period of geologic history is distinct in its geologic and biologic characteristics, and no two units of geologic time were exactly the same (Fig. 1).

The major geologic periods were delineated by 19th-century geologists in Great Britain and western Europe (see Figure 32, p.47). The largest divisions of the geologic record are called eras: they include the Precambrian (time of early life), the Paleozoic (time of ancient life), the Mesozoic (time of middle life), and the Cenozoic (time of recent life). The eras are subdivided into smaller units called periods. Seven periods make up the Paleozoic (though American geologists divide the Carboniferous into Mississippian and Pennsylvanian periods), three consti-

Figure 1 *The geologic time spiral depicting the geologic history of the Earth.*

(Earthquake Information Bulletin 214, courtesy USGS)

tute the Mesozoic, and two make up the Cenozoic. Each period is characterized by somewhat less profound changes in organisms as compared to the eras, which mark boundaries of mass extinctions, proliferations, or rapid transformations of species.

The two periods of the Cenozoic have been further subdivided into seven epochs, which defined the various conditions of the period; for example, the Pleistocene witnessed a series of ice ages. The longest era, the Precambrian, is not subdivided into individual periods as the others are, because the poor fossil content of its rocks provides less detail about the era. Species did not enter the fossil record in large numbers until the beginning of the Paleozoic, about 570 million years ago, when organisms developed hard exterior body parts, probably as a defense against fierce predators. This protection mechanism in turn gave rise to an explosion of life, which resulted in an abundance of fossils.

THE PRECAMBRIAN ERA

The first 4 billion years, or about nine-tenths of geologic time, constitutes the Precambrian, the longest and least understood era of Earth history because of the scarcity of fossils. The Precambrian is subdivided into the Hadean or Azoic eon (time of prelife), 4.6 to 4.0 billion years ago; the Archean eon (time of initial life), 4.0 to 2.5 billion years ago; and the Proterozoic eon (time of earliest life), 2.5 to 0.6 billion years ago. The boundary between the Archean and Proterozoic is somewhat arbitrary and reflects major differences in the characteristics of rocks older than 2.5 billion years and those younger than 2.5 billion years. Archean rocks are products of rapid crustal formation, whereas Proterozoic rocks represent a period of more stable geologic processes.

The Archean was a time when the Earth's interior was hotter, the crust was thinner and therefore more unstable, and crustal plates were more mobile. The Earth was in a great upheaval and subjected to extensive volcanism and meteorite bombardment, which probably had a major effect on the development of life early in the planet's history.

About 4 billion years ago, a permanent crust began to form, composed of a thin layer of basalt embedded with scattered blocks of granite known as rockbergs. Ancient metamorphosed granite in the Great Slave region of Canada's Northwest Territories called Acasta Gneiss indicates that substantial continental crust, comprising as much as one-fifth the present landmass, had formed by this time. The metamorphosed marine sediments of the Isua Formation in a remote mountainous region in southwest Greenland suggest the presence of a saltwater ocean by at least 3.8 billion years ago.

The earliest granites combined into stable bodies of basement rock, upon which all other rocks were deposited. The basement rocks formed the nuclei of the continents and are presently exposed in broad, low-lying dome-like structures called shields (Fig. 2). Precambrian shields are extensive uplifted areas surrounded by sediment-covered bedrock, the continental platforms, which are broad, shallow depressions of basement complex (crystalline rock) filled with nearly flat-lying sedimentary rocks.

Dispersed among and around the shields are greenstone belts, which occupy the ancient cores of the continents. They comprise a jumble of metamorphosed (recrystallized) marine sediments and lava flows caught between two colliding continents. The greenstone belts cover an area of several hundred square miles and are surrounded by immense expanses of gneiss (pronounced like the word *nice*), the metamorphic equivalent of granite and the predominant Archean rock type. The rocks are tinted green as a result of the presence of the mineral chlorite and are among the best evidence for plate tectonics (Fig. 3), the shifting of crustal plates on the Earth's surface, early in the Precambrian.

Figure 2 *The location of Precambrian continental shields, the nuclei upon which the continents grew, which comprise the oldest rocks on Earth.*

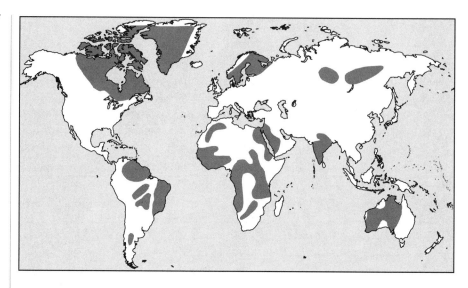

Geologists are particularly interested in greenstone belts because they not only provide important evidence for plate tectonics but also contain most of the world's gold. India's Kolar greenstone belt holds the richest gold deposits. It is some 3 miles wide and 50 miles long and formed when two plates clashed about 2.5 billion years ago. In Africa, the best deposits are in rocks as old as 3.4 billion years, and most South African gold mines are found in greenstone belts. In North America, the best gold mines are in the greenstone belts of the Great Slave region of northwestern Canada, where well over 1,000 deposits are known.

The greenstone belts also comprised ophiolites, from the Greek *ophis,* meaning "serpent." They are slices of ocean floor shoved up onto the conti-

Figure 3 *The plate tectonics model, in which new oceanic crust is generated at spreading ridges and old oceanic crust is destroyed in subduction zones, or trenches, along the edges of continents or island arcs, processes that move the continents around the face of the Earth.*

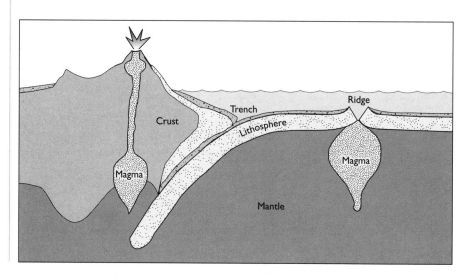

nents by drifting plates and are as much as 3.6 billion years old. In addition, a number of ophiolites contain ore-bearing rocks that are important mineral resources the world over. Pillow lavas, which are tubular bodies of basalt extruded undersea, appear in the greenstone belts as well, signifying that the volcanic eruptions took place on the ocean floor. Because greenstone belts are essentially Archean in age, their disappearance from the geologic record around 2.5 billion years ago marks the end of the Archean eon.

The Proterozoic eon, 2.5 to 0.6 billion years ago, was a shift to calmer times as the Earth matured from adolescence to adulthood. When the eon began, as much as 75 percent of the current continental crust had formed. Continents were more stable and welded together into a single large super-continent. Extensive volcanic activity, magmatic intrusions, and rifting and patching of the crust built up the continental interiors, while erosion and sedimentation built the continental margins outward. The global climate of the Proterozoic was significantly cooler, and the Earth experienced its first major ice age between 2.3 and 2.4 billion years ago (Table 1).

By the beginning of the Proterozoic, most of the material that is now locked up in sedimentary rocks was at or near the surface. In addition, ample sources of Archean rocks were available for erosion and redeposition into

TABLE 1 CHRONOLOGY OF THE MAJOR ICE AGES

Time (years)	Event
10,000–present	Present interglacial
15,000–10,000	Melting of ice sheets
20,000–18,000	Last glacial maximum
100,000	Most recent glacial episode
1 million	First major interglacial
3 million	First glacial episode in Northern Hemisphere
4 million	Ice covers Greenland and the Arctic Ocean
15 million	Second major glacial episode in Antarctica
30 million	First major glacial episode in Antarctica
65 million	Climate deteriorates; poles become much colder
250–65 million	Interval of warm and relatively uniform climate
250 million	The great Permian ice age
700 million	The great Precambrian ice age
2.4 billion	First major ice age

Figure 4 *Glacial land-scape high on the south flank of Uinta Mountains, Duchesne County, Utah. An unnamed ice-sculptured peak at the head of Rock Creek Basin looms above a morainal ridge in the foreground.*

(Photo by W. R. Hansen, courtesy USGS)

Proterozoic rock types. Sediments derived directly from primary sources are called wackes, often described as dirty sandstone. Most Proterozoic wackes composed of sandstones and siltstones originated from Archean greenstones. Another common rock type was a fine-grained metamorphosed rock called quartzite, derived from the erosion of siliceous grainy rocks such as granite and arkose, a coarse-grain sandstone with abundant feldspar.

Conglomerates, which are consolidated equivalents of gravels, were also abundant in the Proterozoic. Nearly 20,000 feet of Proterozoic sediments lie in the Uinta Range of Utah (Fig. 4), one of the only two major east-west trending mountain ranges in North America. The Montana Proterozoic belt system contains sediments over 11 miles thick. The Proterozoic is also known for its terrestrial redbeds, composed of sandstones and shales cemented by iron oxide, which colored the rocks red. Their appearance, around 1 billion years ago, indicates that the atmosphere contained substantial amounts of oxygen, which oxidized the iron in a process similar to the rusting of steel.

The weathering of primary, or parental, rocks during the Proterozoic also produced solutions of calcium carbonate, magnesium carbonate, calcium sulfate, and sodium chloride, which in turn precipitated into limestone, dolomite, gypsum (see Glossary), and halite (rock salt). The Mackenzie Mountains of northwestern Canada contain dolomite deposits more than a mile thick. These minerals are thought to be mainly chemical precipitates and not of biologic origin. Carbonate rocks, such as limestone and chalk, produced chiefly by the deterioration of shells and skeletons of simple organisms, became much more common during the latter part of the Proterozoic, between about 700 and 570 million years ago, whereas during the Archean, they were relatively rare because of the scarcity of lime-secreting organisms.

During the Proterozoic, the continents were composed of odds and ends of Archean cratons, which are ancient, stable rocks in continental interiors. The original cratons formed within the first 1.5 billion years and totaled about one-tenth of the present landmass. They numbered in the dozens and ranged in size from about a fifth the area of today's North America to smaller than the state of Texas. The cratons are composed of highly altered granite and metamorphosed marine sediments and lava flows. The rocks originated from intrusions of magma into the primitive oceanic crust.

Several cratons welded together to form an ancestral North American continent called Laurentia (Fig. 5). Most of the continent, comprising the interior of North America, Greenland, and northern Europe, evolved in a relatively brief period of only 150 million years. Laurentia continued to grow by garnering bits and pieces of continents and chains of young volcanic islands. A major part of the continental crust underlying the United States from Arizona to the Great Lakes to Alabama formed in one great surge of crustal generation around 1.8 billion years ago that has no equal. This buildup possibly resulted from greater tectonic activity and crustal generation during the Proterozoic than during any subsequent time of Earth history.

After the rapid continent building, the interior of Laurentia experienced extensive igneous activity that lasted from 1.6 to 1.3 billion years ago. A broad belt of red granites and rhyolites, which are igneous rocks formed by solidifying of molten magma below ground as well as on the surface, extended sev-

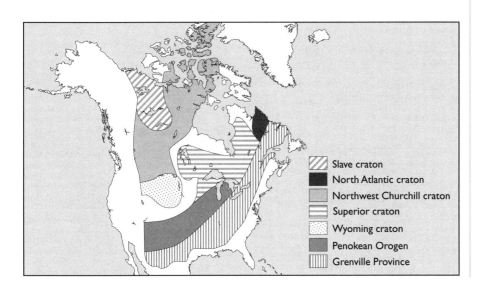

Figure 5 The cratons that constitute North America came together some 2 billion years ago.

Slave craton
North Atlantic craton
Northwest Churchill craton
Superior craton
Wyoming craton
Penokean Orogen
Grenville Province

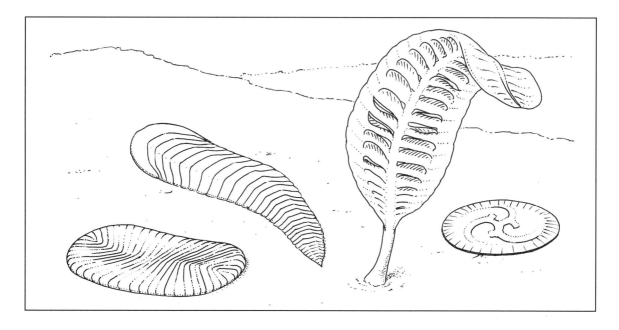

Figure 6 Late
Precambrian Ediacara
fauna from Australia.

eral thousand miles across the interior of the continent from southern California to Labrador. The Laurentian granites and rhyolites are unique because of their sheer volume, which suggests that the continent stretched and thinned almost to the breaking point.

These rocks are presently exposed in Missouri, Oklahoma, and a few other localities but are buried under sediments up to a mile thick in the center of the continent. In addition, vast quantities of molten basalt poured from a huge tear in the crust running from southeast Nebraska into the Lake Superior region about 1.1 billion years ago. Arcs of volcanic rock also weave through central and eastern Canada down into the Dakotas.

Marine life during the Proterozoic was highly distinct from that of the Archean and represented considerable biologic advancement. Numerous impressions of strange extinct species have been found in the Ediacara Formation of southern Australia, dated around 670 million years old (Fig. 6). This great diversity of species followed the Precambrian ice age, the most extensive glaciation on Earth, when nearly half the land surface was covered with glaciers. When the ice retreated and the seas began to warm, life took off in all directions.

Unique and bizarre creatures preserved in Australia's Ediacara Formation thrived in the ocean, and a greater percentage of experimental organisms, animals that evolved unusual characteristics, came into being at that time than during any other interval of Earth history. As many as 100 phyla, organisms that shared similar body styles, existed, whereas only about a third as many phyla are living today. This biologic exuberance set the stage for the

Phanerozoic eon, or the time of later life, comprising the Paleozoic, Mesozoic, and Cenozoic eras. For the first time, fossilized remains of animals became abundant, because of the evolution of lime-secreting organisms that constructed hard shells in the lower Paleozoic.

Toward the end of the Proterozoic, between 630 and 560 million years ago, a supercontinent named *Rodinia,* Russian for "motherland," located near the equator, rifted apart into four or five major continents, although they were configured much differently than they are today. The breakup produced extensive continental margins, where vast carbonate belts formed. This extended shoreline provided additional habitat area, which along with warm Cambrian seas might have played a major role in the rapid explosion of new species by the start of the Paleozoic.

THE PALEOZOIC ERA

The Paleozoic era, which spans a period from about 570 to about 250 million years ago, was a time of intense growth and competition in the ocean and later on the land, culminating with widely dispersed and diversified species. By the middle of the era, all major animal and plant phyla were already in existence. The earliest period of the Paleozoic is called the Cambrian, named for the Cambrian mountain range of central Wales, where sediments containing the earliest known fossils were found. Thus, the base of the Cambrian was once thought to be the beginning of life, and all previous time was known as the Precambrian.

The Paleozoic is generally divided into two time units of nearly equal duration. The lower Paleozoic consists of the Cambrian, Ordovician, and Silurian periods, and the upper Paleozoic comprises the Devonian, Carboniferous, and Permian periods. The first half of the Paleozoic was relatively quiet in terms of geologic processes, with little mountain building, volcanic activity, or glaciation and no extremes in climate. Most of the continents were located near the equator; that location explains the presence of warm Cambrian seas. Sea levels rose and flooded large portions of the land. The extended shoreline might have spurred the explosion of new species, producing twice as many phyla during the Cambrian as before or since. Never were so many experimental organisms in existence, none of which has any modern counterparts. Most new species of the early Cambrian were short-lived, however, and became extinct.

During the late Precambrian and early Cambrian, a proto–Atlantic Ocean called the Iapetus opened, forming extensive inland seas. The inundation submerged most of Laurentia and the ancient European continent called Baltica. The Iapetus Sea was similar in size to the North Atlantic and occupied the same general location about 500 million years ago. It was dotted with volcanic islands, resembling the present-day southwestern Pacific Ocean. The shallow waters of the near-shore environment of this ancient sea contained abundant invertebrates, including trilobites, which accounted for about 70 percent of all species.

During the Cambrian, continental motions assembled the present continents of Africa, South America, Australia, Antarctica, and India into a large landmass called Gondwana (Fig. 7), named for an ancient region of east-central India. Much of Gondwana was in the South Polar region from the Cambrian to the Silurian. The present continent of Australia sat on the equator at the northern edge of Gondwana. A major mountain building episode from the Cambrian to the middle Ordovician deformed areas between all continents comprising Gondwana, indicating their collision during this interval. Extensive igneous activity and metamorphism accompanied the mountain building at its climax.

During the late Silurian, Laurentia collided with Baltica and closed off the Iapetus. The collision fused the two continents into Laurasia, named for the Laurentian province of Canada and the Eurasian continent, about 400 million years ago. These Paleozoic continental collisions raised huge masses of rocks into several mountain belts throughout the world. The sutures joining the landmasses are preserved as eroded cores of ancient mountains called orogens from the Greek word *oros,* meaning "mountain."

Figure 7 *The configuration of the southern continents that comprised Gondwana.*

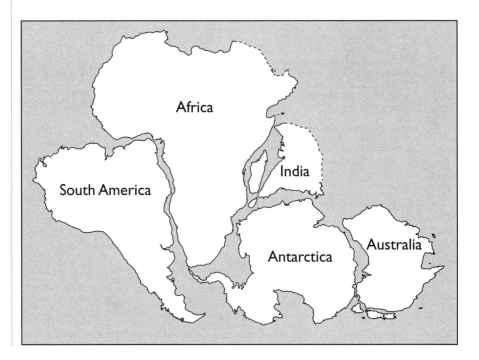

Laurasia and Gondwana were separated by a large body of water, the Tethys Sea, named for the mother of the seas in Greek mythology. Thick deposits of sediments washed off the continents flowed into the Tethys and were later squeezed by continental collisions and uplifted into mountain belts. The continents were lowered by erosion, and shallow seas flowed inland, flooding more than half the landmass. The inland seas and wide continental margins along with a stable environment provided excellent growing conditions for marine life to flourish and migrate throughout the world.

The widespread distribution of evaporite deposits in the Northern Hemisphere, coal deposits in the Canadian Arctic, and carbonate reefs suggest a warm climate and desert conditions over large areas. Warm temperatures of the past are generally indicated by abundant marine limestones, dolostones, and calcareous shales. A coal belt, extending from northeastern Alaska across the Canadian archipelago to northernmost Russia, suggests that vast swamps were prevalent in these regions. The ideal climate setting helped spur the rise of the amphibians that inhabited the great Carboniferous swamps.

The second half of the Paleozoic followed on the heels of a Silurian ice age, when Gondwana wandered into the South Polar region around 400 million years ago and acquired a thick sheet of ice. As the seas lowered and the continents rose, the inland seas departed and were replaced by great swamps. In these regions, vast coal deposits accumulated during the Carboniferous, which had the highest organic burial rates of any period in Earth history. Extensive forests and swamps grew successively on top of one another and continued to add to thick deposits of peat, which were buried under layers of sediment and compressed into lignite, bituminous, and anthracite coal.

Beginning in the late Devonian and continuing into the Carboniferous, Gondwana and Laurasia converged into the supercontinent Pangaea, Greek for "all lands," which comprised some 40 percent of the Earth's total surface area and extended practically from pole to pole. A single great ocean called Panthalassa, Greek for "universal sea," stretched uninterrupted across the rest of the planet. Over the ensuing time, smaller parcels of land continued to collide with the supercontinent until it peaked in size by the beginning of the Triassic, about 210 million years ago.

The closing of the Tethys Sea during the assembly of Pangaea eliminated a major barrier to the migration of species from one continent to another, allowing them to disperse to all parts of the world. Plant and animal life witnessed a great diversity in the ocean as well as on land. A continuous shallow-water margin extended around the entire perimeter of Pangaea, with no major physical barriers to hamper the dispersal of marine life. The formation of Pangaea spurred a great proliferation of plant and animal life and marked a major turning point in evolution of species, during which the reptiles emerged as the dominant land animals.

The Pangaean climate was one of extremes, with the northern and southern regions as cold as the Arctic and the interior as hot as a desert, where almost nothing grew. The massing of continents together created an overall climate that was hotter, drier, and more seasonal than at any other time in geologic history. As the continents rose higher and the ocean basins dropped lower, the land became dryer and the climate grew colder, especially in the southernmost lands, which were covered with glacial ice. Continental margins were less extensive and were narrower, placing severe restrictions on marine habitat. By the close of the Paleozoic, the southern continents were in the grips of a major ice age.

During the Permian, all the interior seas retreated from the land, as an abundance of terrestrial redbeds and large deposits of gypsum and salt were laid down. Extensive mountain building raised massive chunks of crust. A continuous, narrow continental margin surrounded the supercontinent, reducing the shoreline, thus radically limiting the marine habitat area. Moreover, unstable near-shore conditions resulted in an unreliable food supply. Many species unable to cope with the limited living space and food supply died out in tragically large numbers. The extinction was particularly devastating to Permian marine fauna. Half the families of aquatic organisms, 75 percent of the amphibian families, and over 80 percent of the reptilian families, representing more than 95 percent of all known species, abruptly disappeared. In effect, the extinction left the world almost as devoid of species at the end of Paleozoic as at the beginning.

THE MESOZOIC ERA

The Mesozoic era, from about 250 to about 65 million years ago, comprises the Triassic, Jurassic, and Cretaceous periods. When the era began, the Earth was recovering from a major ice age and the worst extinction event in geologic history. Thus, the bottom of the Mesozoic was a sort of rebirth of life, and 450 new families of organisms came into existence. However, instead of developing entirely new body styles, as in the early Paleozoic, the start of the Mesozoic saw only new variations on already established themes. Therefore, fewer experimental organisms evolved, and many of the lines of today's species came into being.

At the beginning of the era, all the continents were consolidated into a supercontinent, about midway they began to break up, and at the end they were well along the path to their present locations (Fig. 8). The breakup of Pangaea created three major bodies of water, the Atlantic, Arctic, and Indian Oceans. The climate was exceptionally mild for an unusually long period, possibly as a result of increased volcanic activity and the resultant greenhouse effect. One group of animals that excelled during these extraordinary condi-

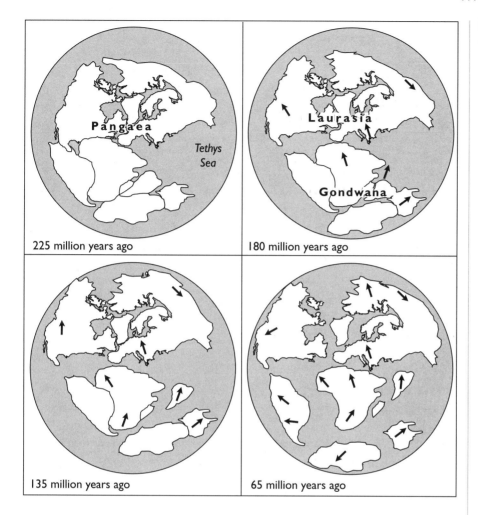

225 million years ago

180 million years ago

135 million years ago

65 million years ago

Figure 8 *The breakup and drift of the continents.*

tions were the reptiles. Some reptilian species returned to the sea; others took to the air. They occupied nearly every corner of the globe; that is why the era is generally known as the "age of the reptiles."

In the early Triassic, the great glaciers of the previous ice age melted, and the seas began to warm. The energetic climate facilitated the erosion of the high mountain ranges of North America and Europe. Seas retreated from the continents as they continued to rise, and widespread deserts covered the land. Abundant terrestrial redbeds and thick beds of gypsum and salt were deposited in the abandoned basins. A preponderance of red rocks composed of sandstones and shales are exposed in the mountains and canyons in the western United States (Fig. 9). Terrestrial redbeds covered a region from Nova Scotia to South Carolina and the Colorado Plateau. Redbeds were also common in Europe, especially in northwestern England.

Figure 9 *A redbed formation on the east side of the Bighorn Mountains, Johnson County, Wyoming.*

(Photo by N. H. Darton, courtesy USGS)

Huge lava flows and granitic intrusions invaded Siberia, and extensive lava flows covered South America, Africa, and Antarctica as well. In South America, great floods of basalt, upward of 2,000 feet or more thick, blanketed large parts of Brazil and Argentina. Triassic basalts in eastern North America erupted from a great rift that separated the continent from Eurasia. Basalt flows also envelop a region from Alaska to California. These large volcanic eruptions created a series of overlapping lava flows, giving many exposures a terracelike appearance known as traps, from the Dutch word for "stairs."

Early in the Jurassic period, North America separated from South America, and a great rift divided the North American and Eurasian continents. The rupture separating the continents flooded with seawater to form the infant North Atlantic Ocean. India, nestled between Africa and Antarctica, drifted away from Gondwana, and Antarctica, still attached to Australia, swung away from Africa to the southeast, forming the proto–Indian Ocean. During the Jurassic and continuing into the Cretaceous, an interior sea flowed into the west-central portions of North America. Massive accumulations of marine sediments eroded from the Cordilleran highlands to the west (sometimes referred to as the ancestral Rockies) were deposited on the terrestrial redbeds of the Colorado Plateau, forming the Jurassic Morrison Formation, well known for fossil bones of large dinosaurs (Fig. 10). Eastern Mexico, southern Texas, and Louisiana were also flooded, and seas inundated South America, Africa, and Australia as well.

The continents were flatter, mountain ranges were lower, and sea levels were higher. Thick deposits of sediment that filled the inland marine basins of North America were uplifted and eroded, providing the western United States with its impressive landscapes. Reef building was intense in the Tethys Sea, and thick deposits of limestone and dolomite were laid down by lime-secreting organisms in the interior seas of Europe and Asia. These deposits were later uplifted during one of geologic history's greatest mountain building episodes. The rim of the Pacific Basin became a hotbed of geologic activity, and practically all mountain ranges facing the Pacific Ocean and island arcs along its perimeter developed during this period.

During the Cretaceous period, plants and animals were especially prolific and ranged practically from pole to pole. Huge deposits of limestone and chalk created in Europe and Asia gave the period its name, from the Latin *creta*, meaning "chalk." Mountains were lower and sea levels higher, and the total land surface declined to perhaps half its present size.

In the late Cretaceous and early Tertiary, land areas were inundated by the ocean, which flooded continental margins and formed great inland seas, some of which split continents in two. Seas divided North America in the Rocky Mountain and high plains regions, South America was cut in two in the region that later became the Amazon Basin, and Eurasia was split by the joining of the Tethys Sea and the newly formed Arctic Ocean. The oceans of the Cretaceous were also interconnected in the equatorial regions by the Tethys and Central American seaways, providing a unique circumglobal oceanic current system that made the climate equable, with no extremes in weather.

Toward the end of the Cretaceous, North America and Europe were no longer in contact except for a land bridge created by Greenland to the north. The Bering Strait between Alaska and Asia narrowed, creating the Arctic

Figure 10 *A dinosaur boneyard at the Howe Ranch quarry near Cloverly, Wyoming. The dinosaurs, along with 70 percent of all other known species, abruptly went extinct 65 million years ago.*

(Photo by N. H. Darton, courtesy USGS)

Ocean, which was practically land-locked. Africa moved northward and began to close the Tethys Sea, leaving Antarctica, which was still attached to Australia, far behind. As Antarctica and Australia moved eastward, a rift developed and began to separate them.

Meanwhile, India began to cross the equator and narrow the gap separating it from southern Asia. The crust rifted open on the west side of India, and massive amounts of molten rock poured onto the landmass, blanketing much of west central India, known as the Deccan Traps. Over a period of several million years about 100 individual basalt flows produced over 350,000 cubic miles of lava, up to 8,000 feet thick. Continental rifting during the same time began separating Greenland from Norway and North America. The rifting poured out great flood basalts across eastern Greenland, northwestern Britain, northern Ireland, and the Faeroe Islands between Britain and Iceland.

When the Cretaceous came to an end, the seas receded from the land as sea levels lowered and the climate grew colder. The decreasing global temperatures and increasing seasonal variation in the weather made the world more stormy, with powerful gusty winds that wreaked havoc over the Earth. These conditions might have had a major impact on the climatic and ecologic stability of the planet, possibly leading to the great extinction at the end of the era.

THE CENOZOIC ERA

The Cenozoic era, from about 65 million years ago to the present, comprises the Tertiary period, which occupies most of the era, and the Quaternary period, which covers the last 2 million years. Both terms were adapted from the old geologic time scale in which the Primary and Secondary periods represented ancient Earth history. The pronounced unequal lengths of the two periods acknowledge a unique sequence of ice ages during the Pleistocene epoch.

Most European and many American geologists prefer to subdivide the Cenozoic into two nearly equal time intervals. The first is the Paleogene period, from about 65 to about 26 million years ago, which includes the Paleocene, Eocene, and Oligocene epochs. The second is the Neogene period, from about 26 million years ago to the present, which includes the Miocene, Pliocene, Pleistocene, and Holocene epochs. Whichever time scale is used, the Cenozoic is generally regarded as the "age of mammals."

The Cenozoic was a time of constant change, as all species had to adapt to a wide range of living conditions. Changing climate patterns resulted from the movement of continents toward their present positions, and intense tectonic activity built a large variety of landforms and raised most mountain ranges of the world. Except for a few land bridges exposed from time to time, plants and animals were prevented from migrating from one continent to another.

About 57 million years ago, Greenland began to separate from North America and Eurasia. Prior to about 4 million years ago, Greenland was largely ice-free, but today the world's largest island is buried under a sheet of ice up to two miles thick. At times, Alaska connected with east Siberia to close off the Arctic Basin from warm water currents originating from the tropics, resulting in the formation of pack ice in the Arctic Ocean.

Antarctica and Australia broke away from South America and moved eastward. The two continents then rifted apart, with Antarctica moving toward the South Pole and Australia continuing in a northeastward direction. In the Eocene, about 40 million years ago, Antarctica drifted over the South Pole and acquired a permanent ice sheet that buried most of its terrain features (Fig. 11).

The Cenozoic is also known for its intense mountain building, when highly active tectonic forces established the geological provinces of the western United States (Fig. 12). The Rocky Mountains, extending from Mexico to Canada, heaved upward during the Laramide orogeny (mountain-building episode) from about 80 million to 40 million years ago. A large number of parallel faults sliced through the Basin and Range Province, between the Sierra Nevada and the Wasatch Mountains, during the Oligocene, producing parallel, north-south–trending mountain ranges. During the last 10 million years, California's Sierra Nevada rose about 7,000 feet.

Figure 11 Taylor Glacier region, Victoria Land, Antarctica.

(Photo by W. B. Hamilton, courtesy USGS)

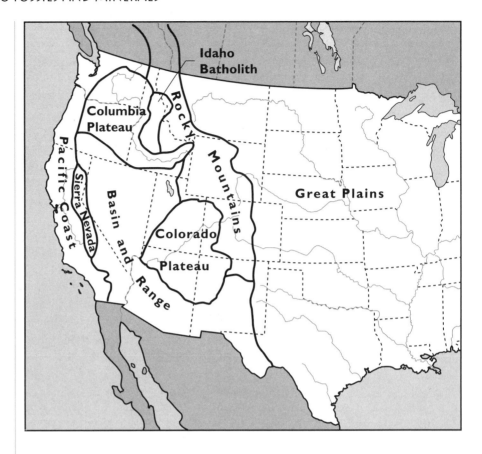

Figure 12 *Geologic provinces of the western United States.*

About 50 million years ago, the collision of the African plate with the Eurasian plate squeezed out the Tethys, creating a long chain of mountains and two major inland seas, the ancestral Mediterranean and a composite of the Black, Caspian, and Aral Seas, called the Paratethys, that covered much of eastern Europe. Thick sediments that had been accumulating for tens of millions of years on the bottom of the Tethys buckled into long belts of mountain ranges on the northern and southern flanks. This episode of mountain building, called the Alpine orogeny, ended around 26 million years ago and marks the boundary between the Paleogene and Neogene periods. The Alps of northern Italy formed when the Italian prong of the African plate thrust into the European plate.

The collision of India with southern Asia, around 45 million years ago, uplifted the tall Himalaya Mountains and the broad three-mile-high Tibetan Plateau, whose equal has not existed on this planet for more than a billion years. The mountainous spine that runs along the western edge of South America forming the Andes Mountains continued to rise throughout more than of the Cenozoic as a result of the subduction of the Nazca plate beneath the South American plate (Fig. 13). The melting of the subducting plate fed magma chambers (volcanic reservoirs) with molten rock, causing numerous volcanoes to erupt in one fiery outburst after another.

Volcanic activity was extensive throughout the world during the Tertiary, whose strong greenhouse effect might explain in part why the Earth grew so warm during the Eocene epoch from 54 million to 37 million years ago. A band of volcanoes stretching from Colorado to Nevada produced a series of very violent eruptions between 30 million and 26 million years ago. The massive outpourings of carbon dioxide–laden lava might have created the extraordinary warm climate that sparked the evolution of the mammals. Winters were warm enough for crocodiles to roam as far north as Wyoming, and forests of palms, cycads, and ferns covered Montana.

Crustal movements in the Oligocene, about 25 million years ago, brought about changes in relative motions between the North American plate and the Pacific plate, creating the San Andreas Fault system running through southern California (Fig. 14). Baja California split off from North America and opened up the Gulf of California. This provided a new outlet to the sea for the Colorado River, which began to carve out the Grand Canyon.

Beginning about 17 million years ago and extending for a period of 2 million years, great outpourings of basalt covered Washington, Oregon, and Idaho, creating the Columbia River Plateau (Fig. 15). Massive floods of lava enveloped an area of about 200,000 square miles, in places reaching 10,000 feet thick. The tall volcanoes of the Cascade Range from northern California to Canada erupted in one great profusion after another. Extensive volcanism

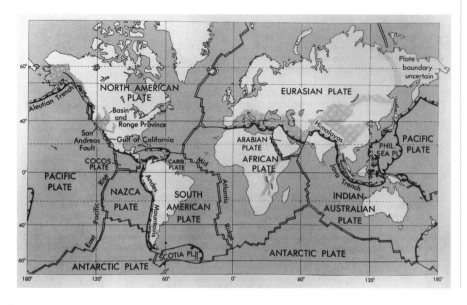

Figure 13 The lithospheric plates that constitute the Earth's crust. Note the position of Nazca and the South American plates.

(Courtesy USGS)

Figure 14 *The San Andreas Fault in southern California.*

(Photo by R. E. Wallace, courtesy USGS)

occurred in the Colorado Plateau and Sierra Madre regions as well. Iceland is an expression of the Mid-Atlantic Ridge, where massive floods of basalt 16 million years ago formed a huge volcanic plateau 900 miles wide, over one-third of which rose above sea level.

About 3 million years ago, the Panama Isthmus separating North and South America uplifted as oceanic plates collided, precipitating a lively exchange of species between the continents. The new landform halted the flow of cold water currents from the Atlantic into the Pacific, which along with the closing of the Arctic Ocean from warm Pacific currents might have initiated the Pleistocene glaciation. Never before has permanent ice existed at both poles, suggesting that the planet has been steadily cooling since the Cretaceous. By the time the continents had wandered to their present positions and the mountain ranges had attained their current elevations, the world was ripe for the coming of the ice age.

Figure 15 *Palouse Falls in Columbia River basalt, Franklin-Whitman Counties, Washington.*

(Photo by F. O. Jones, courtesy USGS)

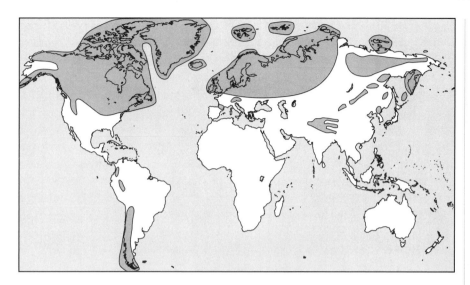

Figure 16 *The extent of glaciation during the last ice age.*

THE PLEISTOCENE ICE AGE

The Pleistocene epoch witnessed a progression of ice ages advancing and retreating almost by clockwork. About 3 million years ago, huge volcanic eruptions in the northern Pacific darkened the skies, and global temperatures plummeted, culminating in a series of glacial episodes. During the last ice age, massive ice sheets swept out of the polar regions, and glaciers up to two miles or more thick enveloped Canada, Greenland, and northern Eurasia (Fig. 16). The glaciers covered some 11 million square miles of land that is presently ice-free. The glaciation began with a rapid buildup of glacial ice some 115,000 years ago, intensified about 75,000 years ago, and peaked about 18,000 years ago.

North America was engulfed by two main glacial centers. The largest glacier, called the Laurentide, blanketed an area of about 5 million square miles. It extended from Hudson Bay and reached northward into the Arctic Ocean and southward into eastern Canada, New England, and the upper midwestern United States. A smaller ice sheet, called the Cordilleran, originated in the Canadian Rockies and enveloped western Canada and the northern and southern sections of Alaska, leaving an ice-free corridor down the center of the present state. Scattered glaciers also covered the mountainous regions of the northwestern United States. Ice buried the mountains of Wyoming, Colorado, and California, and rivers of ice linked the North American cordillera with mountains in Mexico.

Europe was engulfed by two major ice sheets as well. The largest, the Fennoscandian, fanned out from northern Scandinavia and covered most of Great Britain as far south as London and large parts of northern Germany,

Poland, and European Russia. A smaller ice sheet, known as the Alpine and centered in the Swiss Alps, enveloped parts of Austria, Italy, France, and southern Germany. In Asia, glaciers occupied the Himalayas and blanketed parts of Siberia.

In the Southern Hemisphere, only Antarctica held a major ice sheet, which expanded to about 10 percent larger than its present size and extended as far as the tip of South America. Sea ice surrounding Antarctica nearly doubled its modern wintertime area. Smaller glaciers capped the mountains of Australia, New Zealand, and the Andes of South America, the latter of which contained the largest of the southern alpine ice sheets. Throughout the rest of the world, mountain glaciers topped peaks that are currently ice-free.

The lower temperatures reduced the evaporation rate of seawater and decreased the average amount of precipitation, causing expansion of deserts in many parts of the world. The fierce desert winds produced tremendous dust storms, and the dense dust suspended in the atmosphere blocked sunlight, keeping temperatures well below present-day averages. Most of the wind-blown sand deposits called loess in the central United States were laid down during the Pleistocene ice ages.

Approximately 5 percent of the planet's water was locked up in glacial ice. The continental ice sheets contained approximately 10 million cubic miles of water and covered about one-third the land surface with glacial ice three times its current size. The accumulated ice dropped sea levels about 400 feet and shorelines advanced seaward up to 100 miles or more. The drop in sea level exposed land bridges and linked continents, spurring a vigorous migration of species, including humans, to various parts of the world. Adaptations to the cold climate allowed certain species of mammals to thrive in the ice-free regions of the northern lands. Giant mammals, including the mammoth, sabertooth cat, and giant sloth, roamed many parts of the Northern Hemisphere that were free of glaciers.

Perhaps one of the most dramatic climate changes in geologic history took place during the present interglacial known as the Holocene epoch, which began about 11,000 years ago. After some 100,000 years of gradual accumulation of snow and ice up to two miles and more thick, the glaciers melted away in only a matter of several thousand years, retreating several hundred feet annually. The retreating glaciers left an assortment of glacial deposits in their wake, including sinuous eskers, elongated drumlins, and immense boulder fields (Fig. 17). About a third of the ice melted between 16,000 and 12,000 years ago, when average global temperatures increased about five degrees Celsius to nearly present levels. A renewal of the deep-ocean circulation system, which was shut off or weakened during the ice age, might have thawed out the planet from its deep freeze.

The demise of the giant ice sheets and the subsequent warming of the climate left many puzzles such as an unusual occurrence of hippopotamus

Figure 17 *A perched erratic boulder left by the ice of the El Portal glaciation, near the head of Little Cottonwood Creek, east of Army Pass, Inyo County, California.*

(Photo by F. E. Mathes, courtesy USGS)

bones in the deserts of Africa. During a wet period between 12,000 and 6,000 years ago, some of today's African deserts were covered with large lakes. Lake Chad, lying on the border of the Sahara Desert, appears to have swelled over 10 times its present size. Swamps, long since vanished, once harbored large populations of hippopotamuses and crocodiles, whose fossil bones now bake in the desert sands.

After sampling a little geologic history, the next chapter shows how fossils helped uncover clues to the past.

2

CLUES TO THE PAST
THE PRINCIPLES OF PALEONTOLOGY

Fossils have been known from ancient times, and perhaps the first to speculate on their origin were the early Greeks. The Greek philosophers recognized that seashells found in the mountains were the remains of once-living creatures. Although Aristotle clearly recognized that certain fossils such as fish bones were the remains of organisms, he generally believed that fossils were placed in the rocks by a celestial influence. This astrologic account for fossils maintained its popularity throughout the Middle Ages. During this time, competing fossil theories included the idea that fossils grew in rocks, were discarded creations, or were tricks of the devil to deceive humans about the true history of the world. Fossils were also thought to be the creations of Mother Nature in a playful mood.

Not until the Renaissance period and the rebirth of science did people pursue alternate explanations for the existence of fossils that were based on scientific principles. By the 1700s, most scientists began to accept fossils as the remains of organisms because they closely resembled living things rather than merely inorganic substances such as concretions or nodules in rock. When placed in their proper order, fossils pieced together a nearly complete historical account of life on Earth, showing clear evidence for the evolution and extinction of species.

KEYS TO THE HISTORY OF LIFE

Figure 18 *Boiling mud springs northwest of Imperial Junction, California.*

(Photo by W. C. Mendenhall, courtesy USGS)

One of the major problems encountered when exploring for fossils of early life is that the Earth's crust is constantly rearranging itself, and only a few fossil-bearing formations have survived undisturbed over time, the others having been eroded away. Therefore, the history of the Earth as told by its fossil record is not completely known because of the remaking of the surface, which erases whole chapters of geologic history. Yet the study of fossils along with the radiometric dating of the rocks that contain them have constructed a reasonably good chronology of Earth history.

Bacteria, which descended from the earliest known form of life, remain by far the most abundant organisms, and without them no other life-forms could exist. Evidence that life began very early in the Earth's history when the planet was still quite hot exists today in the form of thermophilic (heat-loving) bacteria, found in thermal springs and other hot-water environments (Fig. 18). Because these bacteria lack a nucleus, which ceases to function in hot water, they can live and reproduce successfully even at temperatures well above the normal boiling point of water as long as it remains a liquid, which requires pressures equal to those in the deep sea. The existence of these organisms is used as evidence that thermophiles were the ancestors of all life on Earth.

Life probably had a very difficult time at first. When living organisms began evolving, the Earth was constantly bombarded with large meteorites. As a result, the first living forms might have been repeatedly killed off, forcing life to regenerate over and over again. Whenever primitive organic molecules began to be arranged into living cells, the gigantic impacts would have blasted them apart before they could reproduce. One safe place where life would

Figure 19 *Tall tube worms, giant clams, and large crabs occupy the seafloor near the hydrothermal vents.*

be free to evolve was the bottom of the ocean, where hydrothermal vents provided warmth and nourishment. Today, these areas contain some of the most bizarre creatures the Earth has ever known (Fig. 19).

Among the oldest fossils found on Earth are the remains of ancient microorganisms and stromatolites (Fig. 20), layered structures formed by the accretion of fine sediment grains by matted colonies of cyanobacteria (formerly called blue-green algae). These were found in 3.5-billion-year-old sedimentary rocks of the Warrawoona group in a desolate place called North Pole in Western Australia. Associated with these rocks were cherts (hard rocks composed of microscopic crystals of silica) with microfilaments, which are small, threadlike structures, possibly of bacterial origin.

Most Precambrian cherts are thought to be chemical sediments precipitated from silica-rich water in a deep ocean. The abundance of chert in the early Precambrian is evidence that most of the Earth's crust was deeply submerged in a global ocean during that time. However, cherts at the North Pole site appear to have had a shallow-water origin. This silica probably leached out of volcanic rocks that erupted into shallow seas. The silica-rich water circulated through porous sediments, dissolving the original minerals and precipitating silica in their place. Microorganisms buried in the sediments were encased

Figure 20 *Stromatolite beds from a cliff above the Regal mine, Gila County, Arizona.*

(Photo by A.F. Shride, courtesy USGS)

in one of the hardest natural substances and thus were able to withstand the rigors of time.

Similar cherts with microfossils of filamentous bacteria dating 3.4 billion years old have been found in eastern Transvaal, South Africa. In addition, 2-billion-year-old cherts from the Gunflint iron formation on the north shore of Lake Superior contained similar microfossils. These rocks were originally mined for flint to fire the flintlock rifles of the early settlers until the discovery there of iron, which made this region one of the best iron mining districts in the country.

About 500 million years after the formation of the Gunflint chert, a new type of cell, called a eukaryote, emerged in the fossil record. It was characterized by a nucleus that allowed chromosomes to divide and unite hereditary material in a systematic manner. A greater number of genetic mutations were produced, providing a wide variety of organisms, some of which might have adapted to their environment better than others. These organisms were the forerunners of all the complex forms of life on Earth today.

By far, the most numerous fossils representing the first abundant life on Earth were the hard parts of marine animals lacking backbones called invertebrates. Perhaps the best known of these creatures was the trilobite (Fig. 21),

Figure 21 *Trilobite fossils of the Cambrian age Carrara Formation in the southern Great Basin of California and Nevada.*

(Photo by A. R. Palmer, courtesy USGS)

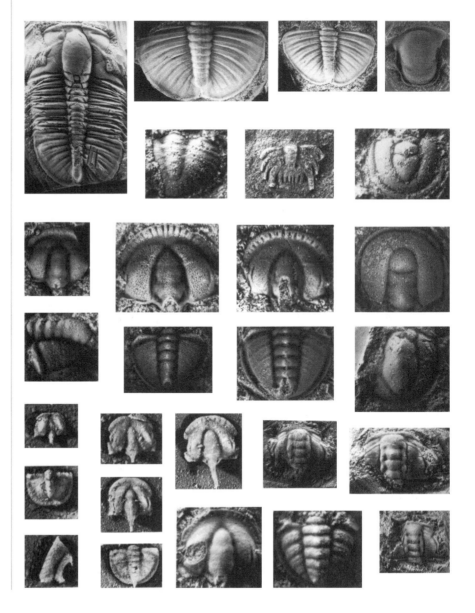

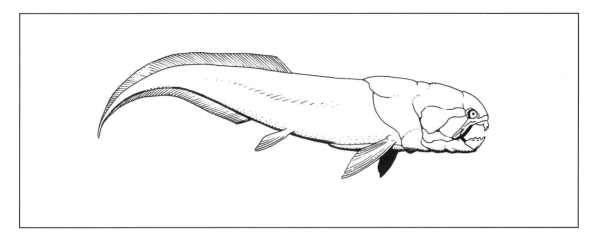

a primitive arthropod and ancestor of today's horseshoe crab. They first appeared at the base of the Paleozoic era, about 570 million years ago. The trilobites became the dominant animals of the Paleozoic, diversifying into some 10,000 species before declining and becoming extinct after some 340 million years of existence. Because trilobites were so widespread and lived for so long, their fossils have become important markers (also called guide or index fossils) for dating Paleozoic rocks.

The demise of the trilobites might be connected to the arrival of the jawed fishes. Fish were among the first vertebrates, or animals with internal skeletons. These provided more efficient muscle attachments, which gave fish much better mobility than their invertebrate counterparts. Fish constitute more than half the species of vertebrates, both living and extinct. The placoderms (Fig. 22) were fierce giants, growing 30 feet in length. They had thick armor plating around the head that extended over and behind the jaws and probably made them poor swimmers. They might have preyed on smaller fish, which in turn fed on trilobites.

While fish were thriving in the ocean, plants advanced onto the land beginning some 450 million years ago (Fig. 23). Within 90 million years, vast forests covered the Earth. Their decay, burial, and metamorphism formed many of today's coal deposits (fossil fuel). Evolving along with the land plants were the arthropods, which constitute the largest phylum of living organisms and number roughly 1 million species, or about 80 percent of all known animals. These insects helped to pollinate the plants, whose flowers offered sweet nectar in return for services rendered. Unfortunately, because of their delicate bodies, insects did not fossilize well. However, they could be preserved if trapped in tree sap, which later hardened into amber, a clear yellow substance that allows the study of even the most minute body parts.

The vertebrates did not set foot on dry land until nearly 100 million years after the plants appeared. The first to come ashore were the amphib-

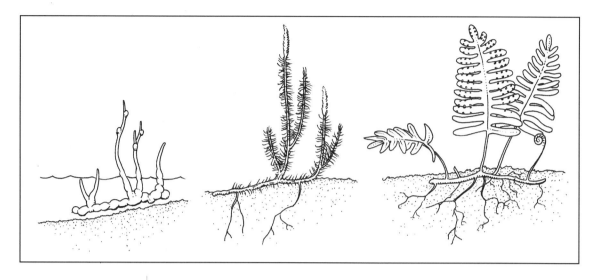

Figure 23 *The emergence of plants from the sea onto the land.*

ians, which evolved into reptiles, which in turn gave rise to the dinosaurs. Dinosaur bones are abundant in Jurassic and Cretaceous sediments in many parts of the United States, particularly in the West. Alongside the dinosaurs evolved the mammals, which for the most part were small nocturnal creatures that fed during the night so as not to compete directly with the dinosaurs. The Cenozoic mammals are well represented in geologic history. Woolly mammoths (Fig. 24), extinct giant mammals of the late Pleistocene ice age, have been well preserved in the deep freeze on top of the world.

EVIDENCE FOR EVOLUTION

During a five-year period from 1831 to 1836, the British naturalist Charles Darwin was employed as the ship's geologist aboard the H.M.S. *Beagle* and described in great detail the rocks and fossils he encountered on his journey around the world (Fig. 25). Darwin was trained as a geologist and thought like one, but today he tends to be viewed as a biologist. He made many significant contributions to the field of geology, which during his day was entering its golden age.

When Darwin visited the Galápagos Islands in the eastern Pacific, he noticed major differences between plants and animals living on the islands and their relatives on the adjacent South American continent. Animals such as finches and iguanas assumed forms that were distinct from but related to those of animals on adjacent islands. Cool ocean currents and volcanic rock made the Galapagos a much different environment than Ecuador, the nearest land 600 miles to the east. The similarities among animals of the two regions could

Figure 24 *The woolly mammoth went extinct at the end of the last ice age.*

only mean that Ecuadorian species colonized the islands and then diverged by a natural process of evolution.

Darwin observed the relationships between animals on islands and on adjacent continents as well as between animals and fossils of their extinct relatives. This study led him to conclude that species had been continuously evolving throughout time. Actually, Darwin was not the first to make this observation. His theory differed, however, in postulating that new parts evolved in many tiny stages rather than in discrete jumps, which he attributed to gaps in the geologic record caused by periods of erosion or nondeposition.

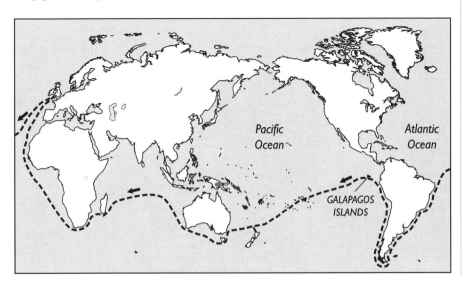

Figure 25 *Darwin's journey around the world. Starting out from Great Britain, he sailed to South America, Australia, Africa, and back to Great Britain.*

Therefore, to Darwin, evolution worked at a constant tempo as species adapted to a constantly changing environment.

Darwin coined the phrase "survival of the fittest," meaning that members of a particular species that can best utilize their environment have the best chance of producing offspring that possess the survival characteristics of their parents. In other words, successful parents have a better chance of passing on their "good" genes to their offspring, which in turn are better able to survive in their respective environment. Natural selection therefore favored those best suited to their environment at the expense of less suited species. Contemporary geologists embraced Darwin's theory, for at last a clear understanding of the changes in body forms in fossils of different ages was at hand. Thus, they could place geologic events in their proper sequence by studying the evolutionary changes that took place among fossils.

Evolution apparently was not always gradual and constant in tempo as Darwin saw it. The fossil record implies that life evolved by fits and starts. Long periods of little or no change were punctuated by short intervals of rapid change and then followed by long periods of stasis (stay-as-is). The pattern of change and stasis is called punctuated equilibrium. Species formed relatively quickly as a result of rapid bursts of evolutionary change. New species evolved within a few thousand years (practically instantaneously in geologic time) and then remained essentially unchanged for up to several million years.

Species formed relatively quickly as a result of rapid bursts of evolutionary change. Furthermore, rapid evolutionary changes in large segments of organisms might appear in the fossil record as having been caused by mass extinction when in fact no actual extinction had occurred. Evolution also might be opportunistic, with variations arising by chance and selected in accordance with the demands of the environment. The evolution of a single species also affects the evolution of others with which it interacts. When the environment changes abruptly to one that is harsher, species incapable of rapidly adapting to these new conditions cannot live at their optimum and therefore do not pass on their "bad" genes to future generations.

Rapid evolutionary advancements might result from rare large mutations. Thus, evolution appears to make sudden leaps, with major changes occurring simultaneously in many body parts. In other words, natural selection does not favor piecemeal tinkering; it therefore cannot work on structures that are not fully functional during intermediary periods of development of new appendages. An example is the development of insect wings, which were probably first used for cooling purposes. Later, as the benefits of flight became apparent, the wing structures grew more aerodynamic, giving flying insects an enormous advantage over their earth-bound competitors.

Evolutionary trends varied throughout geologic time in response to major environmental changes, as natural selection acted to adapt organisms to the new conditions forced on them by several environmental factors, such as chemical alterations in the ocean, climate changes, or mass extinctions. However, natural selection is not deterministic. Variations are purely accidental and selected according to the demands of the environment; the most adapted species have the best chance of survival. Most of the time, species resist change, even though the consequences make them better suited to environmental needs.

Gaps in the fossil record might result from the lack of intermediary species, or so-called missing links, which apparently existed only in small populations. Small populations are less likely to leave a fossil record because the process of fossilization favors large populations. Furthermore, the intermediates probably did not live in the same locality as their ancestors and thus were unlikely to be preserved along with them. New species that start out in small populations evolve rapidly as they radiate into new environments. Then as populations increase, slower evolutionary changes take place as the species' chances of entering the fossil record improve.

The fossil record also might suggest differences in fossil samples where no actual differences exist. In any ecologic community, a few species occur in abundance, some occur frequently, but most are rare and occur only infrequently. In addition, the odds of any individual's becoming fossilized after death, and thus entering the fossil record, are extremely small. No single fossil sample will contain all the rare species in an assemblage of species. If this sample were compared with another higher up in the stratigraphic column, which represents a later time in geologic history, an overlapping but different set of rare species would be recorded. Species found in the lower sample but not in the upper sample might erroneously be inferred to have gone extinct. Conversely, species that appear in the upper sample but not in the lower sample might wrongly be thought to have originated there. Thus, the reading of the fossil record often can be confusing and misleading.

A commonly held belief among scientists is that environmental change drives evolution and not the other way around. However, the British chemist James Lovelock turned the scientific community on its head in 1979 by proposing the Gaia hypothesis, named for the Greek goddess of the Earth. He postulated that the living world is able to control to some extent its own environment and that living organisms maintain the optimal conditions for life by regulating the climate, similarly to the way the human body regulates its temperature to maintain optimal metabolic efficiency. The Gaia hypothesis also suggests that from the very beginning, life followed a well organized pattern of growth independent of chance and natural selection. Apparently, living things kept pace with all the changes in the Earth over time and might have made some major alterations of their own such as converting most of the carbon dioxide in the early atmosphere and ocean into oxygen through photosynthesis (Table 2).

TABLE 2   EVOLUTION OF LIFE AND THE ATMOSPHERE

Evolution	Origin (million years)	Atmosphere
Origin of Earth	4,600	Hydrogen, helium
Origin of life	3,800	Nitrogen, methane, carbon dioxide
Photosynthesis	2,300	Nitrogen, carbon dioxide, oxygen
Eukaryotic cells	1,400	Nitrogen, carbon dioxide, oxygen
Sexual reproduction	1,100	Nitrogen, oxygen, carbon dioxide
Metazoans	700	Nitrogen, oxygen
Land plants	400	Nitrogen, oxygen
Land animals	350	Nitrogen, oxygen
Mammals	200	Nitrogen, oxygen
Humans	2	Nitrogen, oxygen

Perhaps the greatest forces affecting evolutionary changes were plate tectonics and the drifting of the continents. Continental motions had a wide-ranging effect on the distribution, isolation, and evolution of species. The changes in continental configuration greatly affected global temperatures, ocean currents, productivity, and many other factors of fundamental importance to life. The positioning of the continents with respect to each other and to the equator helped determine climatic conditions. When most of the land huddled near the equatorial regions (Fig. 26), the climate was warm, but when lands wandered into the polar regions the climate grew cold and brought periods of glaciation.

The changing shapes of the ocean basins due to the movement of continents affect the flow of ocean currents, the width of continental margins, and, consequently, the abundance of marine habitats. When a supercontinent breaks up, more continental margins are created, the land lowers, and the sea level rises, providing a larger habitat area for marine organisms. During times of highly active continental movements, the Earth experiences greater volcanic activity, especially at spreading centers, where tectonic plates are pulled apart by upwelling magma from the upper mantle. The amount of volcanism could affect the composition of the atmosphere, the rate of mountain building, the climate, and inevitably life itself.

MASS EXTINCTIONS

Practically all species that have ever existed on Earth are extinct (Table 3). Throughout geologic history, species have come and gone on geologic time

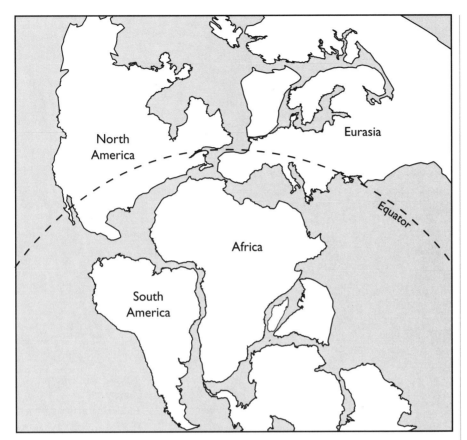

Figure 26 *The approximate positions of the continents relative to the equator during the Devonian and Carboniferous periods.*

scales, so that those living today represent only a tiny fraction of the total. Of the 4 billion species of plants and animals thought to have existed in the geologic past, over 99 percent have become extinct. All extinction events appear to indicate biologic systems in extreme stress brought on by a radical change in the environment caused by large meteorite impacts or volcanic eruptions. Most mass extinctions followed periods of environmental upheavals such as global cooling.

The more devastating and globally encompassing an extinction event, the greater the evolutionary change. For this reason, extinctions play an enormous role in evolution. Extinction is therefore an inevitable part of the evolutionary process and essential for the advancement of species. Therefore, little happens in evolution without extinction's first disrupting living conditions. Each mass extinction marks a watershed in the evolution of life, by resetting the evolutionary clock, forcing species to start anew. When a major extinction event occurs, new species develop to fill vacated habitats. Because of their significant impacts on life, major extinction events also mark the boundaries between geologic periods.

TABLE 3   RADIATION AND EXTINCTION OF SPECIES

Organism	Radiation	Extinction
Mammals	Paleocene	Pleistocene
Reptiles	Permian	Upper Cretaceous
Amphibians	Pennsylvania	Permian-Triassic
Insects	Upper Paleozoic	No major extinction
Land plants	Devonian	Permian
Fishes	Devonian	Pennsylvanian
Crinoids	Ordovician	Upper Permian
Trilobites	Cambrian	Carboniferous and Permian
Ammonoids	Devonian	Upper Cretaceous
Nautiloids	Ordovician	Mississippian
Brachiopods	Ordovician	Devonian and Carboniferous
Graptolites	Ordovician	Silurian and Devonian
Foraminiferans	Silurian	Permian and Triassic
Marine invertebrates	Lower Paleozoic	Permian

The vast majority of the Earth's fauna and flora lived during the Phanerozoic eon from about 570 million years ago (mya) to the present (Fig. 27). This was a period of phenomenal growth as well as tragic episodes of mass extinction, each involving the loss of more than half the species living at the time. Five major extinctions, interspersed with five or more minor die outs, occurred during this period. The first mass extinction occurred in the early Cambrian (530 mya) and decimated over 80 percent of all marine animal genera; it was one of the worst in geologic history. A second mass dying at the end of the Ordovician (440 mya) eliminated some 100 families of marine animals. Another major die off during the middle Devonian (365 mya) witnessed the mass disappearance of many tropical marine groups.

The greatest loss of life in the fossil record occurred at the end of the Permian (250 mya), when half the families of organisms comprising more than 95 percent of all marine species and 80 percent of all terrestrial species, disappeared. Another tragic event at the end of the Triassic (210 mya) took the lives of nearly half the reptilian species. The most familiar die out eliminated 70 percent of all known species including the dinosaurs (Fig. 28) at the end of the Cretaceous (65 mya). All extinction events seem to indicate biologic systems in extreme stress brought on by climate change or a drop in sea level.

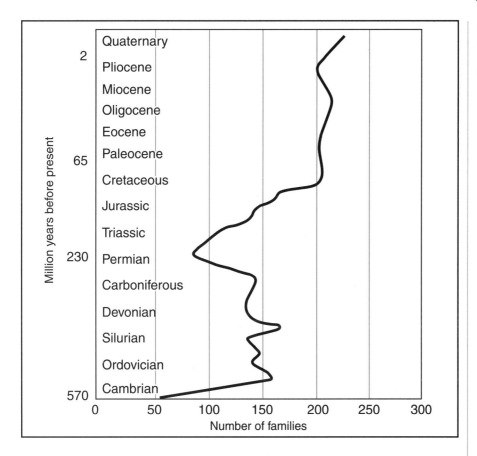

The fossil record suggests that mass extinctions might be somewhat periodic, possibly resulting from celestial influences such as cosmic rays from supernovas or huge meteorite impacts. Ten or more large asteroids or comets have collided with the Earth over the last 600 million years. Analysis of 13 major impact craters distributed over a period from 250 million to 5 million years ago suggests a rate of one crater roughly every 28 million years.

Since the great Permian catastrophe, eight significant extinction events, which defined the boundaries of the geologic time scale, have occurred; many of the strongest peaks have coincided with the boundaries between geologic periods. The episodes of extinction appear to be cyclical, occurring every 26 to 32 million years. Longer intervals of 80 to 90 million years are related to the breakup and collisions of continents. Exceptionally strong extinctions occur every 225 to 275 million years, corresponding to the solar system's rotational period around the center of the Milky Way galaxy.

The extinctions might merely be episodic, with relatively long periods of stability followed by random, short-lived (geologically speaking) extinction events that only appear periodic. Major extinctions therefore could reflect a

Figure 28 *All species of dinosaurs went extinct at the end of the Cretaceous as a result of a number of suspected causes, including terrestrial as well as extraterrestrial influences.*

clustering of several minor events at certain times that, because of the nature of the fossil record, only mimics a cyclical pattern. In other words, random groupings of extinct species on a geologic time scale that is itself uncertain could simply be coincidental. Furthermore, a short period of rapid evolution might manifest itself in the geologic record as though preceded by a mass extinction when indeed none had occurred.

Major extinctions are separated by periods of lower extinction rates, called background extinctions, and the difference between them is only a matter of degree. Therefore, mass extinctions are not simply intensifications of processes operating during background times. Species have regularly come and gone even during optimal conditions. Those that suffer extinction might have been developing certain unfavorable traits. Extinct species also might have lost their competitive edge and been replaced by a superior, more adaptable species. Certain characteristics that permit a species to live successfully during normal periods for some reason become irrelevant when major extinction events occur. Thus, the extinct dinosaurs might not have done anything "wrong" biologically.

The distinction between background and mass extinctions might be distorted by ambiguities in the fossil record, especially when certain species are favored over others for fossilization. Only under demanding geologic conditions that promote rapid burial with little predation or decomposition are the bodies of dead organisms preserved to withstand the rigors of time. Because species with hard body parts fossilize better than soft-bodied organisms, they are more likely to be represented in the fossil record and therefore present a skewed account of historic geology.

Catastrophic extinction events appear to be virtually instantaneous in the fossil record because discerning a period of several thousand years over millions of years of geologic time is not possible. More likely, the extinctions occurred over lengthy periods of perhaps a million years or more, and because of erosion or nondeposition of the sedimentary strata that preserve species as fossils the die outs only appear sudden. Several times in Earth history, sea levels have fallen, reducing sedimentation rates and the preservation of species. Therefore, a sudden break in geologic time might in reality have extended over a lengthy period.

Those species that survive mass extinction radiate outward to fill new environments, which in turn produce entirely new species. These might develop novel adaptations that give them a survival advantage over other species. The adaptations might lead to exotic-looking species that prosper during normal background times, but because of their overspecialization are incapable of surviving mass extinctions. Therefore, the fossil record shows myriad strange creatures, the likes of which have never been seen since (Fig. 29).

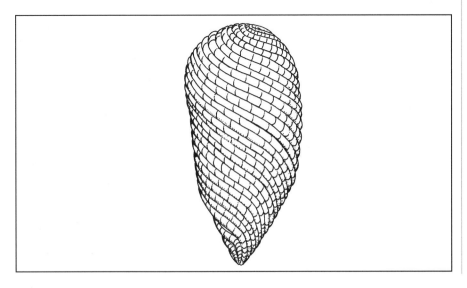

Figure 29 Helicoplacus, an experimental species with body parts assembled differently from those of every other living creature, became extinct about 510 million years ago after surviving for 20 million years.

The geologic record implies that nature is constantly experimenting with new forms of life, and when one fails, such as the dinosaurs, it becomes extinct, and the odds against the reappearance of its particular combination of genes are astronomical. Thus, evolution seems to run on a one-way track, and although it perfects species to live at their optimum in their respective environments, it can never go back to the past. That is why even though the environment of the future might match that of the Cretaceous period when the dinosaurs roamed the Earth, they will never return.

Dinosaurs were not the only ones to go, and 70 percent of all known species vanished at the end of the Cretaceous. Therefore, something in the environment made them all unfit to survive, yet this factor did not significantly affect the mammals. Dinosaurs and mammals coexisted for more than 100 million years. After the dinosaurs became extinct, the mammals underwent an explosive evolutionary radiation, which gave rise to many unusual species, some of which became extinct early in the Cenozoic.

Many geologists are beginning to accept catastrophe as a normal occurrence in Earth history and as a part of the uniformitarian process, also called gradualism. Certain periods of mass extinctions appear to be the result of some catastrophic event, such as the bombardment of one or more large asteroids or comets, rather than subtle changes, such as a change in climate or sea level or an increase in predation. Therefore, mass extinctions appear to be part of a pattern of life throughout the Phanerozoic.

GEOLOGIC AGE DATING

Both large and small extinctions were used by 19th-century geologists to define the boundaries of the geologic time scale (Table 4). But because no means was available to date rocks, the entire geologic record was delineated by using relative dating techniques, which only indicated which bed was older or younger in accordance to its fossil content. Therefore, relative dating only places rocks in their proper sequence but does not indicate how long ago an event took place, only that it followed one event and preceded another. Before the development of radiometric dating techniques, geologists had no method of dating events precisely. So relative dating techniques were developed, and they are still in use today. Absolute dating methods did not replace these techniques, however, but only supplemented them.

The problem with assigning absolute dates to units of relative time is that most radioactive isotopes are restricted to igneous rocks. Even if sedimentary rocks, which constitute most of the rocks on the Earth's surface and contain practically all the fossils, did possess a radioactive mineral, most rocks could not be dated accurately because the sediments were composed of grains derived

TABLE 4   THE GEOLOGIC TIME SCALE

Era	Period	Epoch	Age (millions of years)	First Life Forms	Geology
Cenozoic	Quaternary	Holocene	0.01		
		Pleistocene	2	Humans	Ice age
		Pliocene	11	Mastodons	Cascades
		Neogene*			
		Miocene	26	Saber-tooth tigers	Alps
	Tertiary	Oligocene	37		
		Paleogene*			
		Eocene	54	Whales	
		Paleocene	65	Horses, Alligators	Rockies
Mesozoic	Cretaceous		135		
				Birds	Sierra Nevada
	Jurassic		210	Mammals	Atlantic
				Dinosaurs	
	Triassic		250		
Paleozoic	Permian		280	Reptiles	Appalachians
	Pennsylvanian		310	Trees	Ice age
	Carboniferous				
	Mississippian		345	Amphibians, Insects	Pangaea
	Devonian		400	Sharks	
	Silurian		435	Land plants	Laurasia
	Ordovician		500	Fish	
	Cambrian		570	Sea plants, Shelled animals	Gondwana
			700	Invertebrates	
Proterozoic			2,500	Metazoans	
			3,500	Earliest life	
Archean			4,000		Oldest rocks
			4,600		Meteorites

* See page 16 ("The Cenozoic Era").

from older rocks. Therefore, in order to date sedimentary rocks, geologists had to relate them to igneous masses. A layer of volcanic ash deposited above or below a sedimentary bed could be dated radiometrically, as could cross-cutting features such as granitic dikes which are always younger than the beds they cross. The sedimentary strata would then be bracketed by dated materials, and the age could be estimated.

The radiometric dating method measures the ratios of radioactive parent materials to their daughter products and compares this ratio to the known half-lives of the radioactive elements. The half-life is the time required for one-half of a radioactive element to decay to a stable daughter product. For example, if one pound of a hypothetical radioactive element had a half-life of 1 million years, then after a period of 1 million years a half-pound of the original parent material and a half-pound of daughter product would be present. The ratio of parent element to its daughter product is determined by chemical and radiometric analysis of the sample rock. Therefore, if the quantities of parent and daughter are equal, one half-life has expired, making the sample 1 million years old. After 2 million years, one-quarter of the original parent element remains in the sample, and after 4 million years only one-sixteenth is left. Generally, radioactive elements are usable for age dating up to about 10 half-lives. Afterward, the amount of parent material is reduced to about a thousandth of its original mass.

Radioactive decay also appears to be constant with time and is unaffected by chemical reactions, temperature, pressure, or any other known conditions or processes that could change the decay rate throughout geologic history. Confirmation that decay rates are steady throughout time is found in certain minerals such as biotite mica. Extremely small zones of discoloration, or halos, are found surrounding minute inclusions of radioactive particles within the crystal. The haloes consist of a series of concentric rings around the radioactive source. Particles emitted by the radioactive source damage the surrounding biotite minerals. The energy of the particle is determined by the distance it travels through the mineral and depends on the type of radioactive element responsible. Since the radii of concentric rings corresponds to the energy of present-day particles, particle energies have not changed, and therefore the rate of radioactive decay remains constant over time.

The precision of radiometric age dating depends on the accuracy of the chemical and radiometric analyses that determine the amount of the parent element and daughter product; it also depends on whether either has been added to or removed from the sample since deposition. The quantities of these substances might only be on the order of a few parts per million of the rock mass. A certain amount of naturally occurring daughter material might have existed in the rock before the parent element began decaying. Moreover, many radioactive elements do not decay directly into stable daughter products but go through a series of intermediate decay schemes, further complicating analysis.

Of all the radioactive isotopes that exist in nature, only a few have been proved useful in dating rocks. All others either are very rare or have half-lives that are too short or too long. Rubidium-87 with a half-life of 47 billion years, whose daughter product is strontium-87, is useful for dating rocks older than 20 million years. Uranium-238 with a half-life of 4.5 billion years, and uranium-235 with a half-life of 0.7 billion years, whose daughter products are lead-206 and lead-207, respectively, are useful for dating rocks more than 100 million years old. The uranium isotopes are important for dating igneous and metamorphic rocks. Because both species of uranium occur together, they also can be used to cross-check each other.

Potassium-40 is more versatile for dating younger rocks. Although the half-life of potassium-40 is 1.3 billion years, recent analytical techniques allow the detection of minute amounts of its stable daughter product argon-40 in rocks as young as 30,000 years old. It is less precise for dating younger rocks because of the relatively small amount of daughter product available in the sample. Minerals such as hornblende, nepheline, biotite, and muscovite are used for dating most igneous and metamorphic rocks by the potassium-argon method.

Sedimentary rocks present a more difficult problem for radiometric dating because their material was derived from weathering processes. Fortunately, a micalike mineral called glauconite forms in the sedimentary environment and contains both potassium-40 and rubidium-87. As a result, the age of the sedimentary deposit can be established directly by determining the age of the glauconite. Unfortunately, metamorphism, no matter how slight, might reset the radiometric clock by moving the parent and daughter products elsewhere in the sample. In this case, the radiometric measurement can only date the metamorphic event. In order to date these rocks accurately, a whole-rock analysis must be made, using large chunks of rock instead of individual crystals. Sediments also can be dated by using optically stimulated thermoluminescence, which measures when sand grains were last exposed to light and is especially useful for dating fossil footprints.

To date more recent events, the radioactive isotope carbon-14, or radiocarbon, is used. Carbon-14 is continuously created in the upper atmosphere by cosmic ray bombardment of gases, which in turn release neutrons. The neutrons bombard nitrogen in the air, causing the nucleus to emit a proton, thus converting nitrogen into radioactive carbon-14. In chemical reactions, this isotope behaves identically to natural carbon-12. It reacts with oxygen to form carbon dioxide, circulates in the atmosphere, and is absorbed directly or indirectly by living matter (Fig. 30). As a result, all organisms contain a small amount of carbon-14 in their bodies. Carbon-14 decays at a steady rate with

Figure 30 *The carbon-14 cycle. Cosmic rays striking the atmosphere release neutrons that strike nitrogen atoms to produce carbon-14, which is converted into carbon dioxide and taken in by plants and animals.*

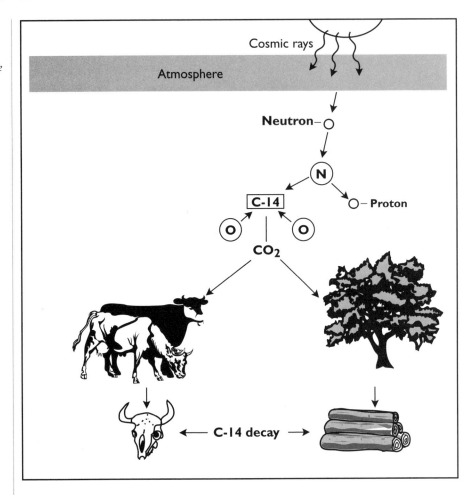

a half-life of 5,730 years. When an organism is alive, the decaying radiocarbon is continuously being replaced, and the proportions of carbon-14 and carbon-12 remain constant. However, when a plant or animal dies, it ceases to take in carbon and the amount of carbon-14 gradually decreases as it decays to stable nitrogen-14. This results from the emission of a beta particle (free electron) from the carbon-14 nucleus, thus transmuting a neutron into a proton and restoring the nitrogen atom to its original state.

Radiocarbon dates are determined by chemical analysis, which compares the proportion of carbon-14 to that of carbon-12 in a sample (Fig. 31). The development of improved analytical techniques has increased the usefulness of radiocarbon dating, and it can be used to date events taking place more than 100,000 years ago. Furthermore, paleontologists, anthropologists, archaeologists, and historians now have a means of accurately dating events from our distant past.

Figure 31 *Scientist dating a sample by the radiocarbon method.*

(Courtesy USGS)

THE GEOLOGIC TIME SCALE

Geologists measure geologic time by tracing fossils through the rock strata and noticing the greater change in the deeper rocks than in those near the surface. Fossil-bearing strata can be followed horizontally over great distances, because a particular fossil bed can be identified in another locality with respect to beds above and below it. These are called marker beds and are used for identifying geologic formations and for delineating rock units for geologic mapping.

When fossils are arranged according to their age, they do not present a random or haphazard picture, but instead show progressive changes from simple to complex forms and reveal the advancement of the species through time. Geologists are thus able to recognize geologic time periods on the basis of groups of organisms that were especially plentiful and characteristic during a particular time. Within each period, many subdivisions are determined by the occurrence of certain species, and this same succession is found on every major continent and is never out of order.

Fossils are necessary for correlating rock units over vast distances because the lithology of rock strata changes from one location to another. A major problem is that few fossil-bearing formations have survived undisturbed over time. Since certain species have existed only during specific intervals, their respective fossils can place stratigraphic units in their proper sequence, which defines the relative time periods. These beds can then be traced over wide areas by comparing their fossil content, providing a comprehensive geologic history over a broad region. The use of fossils has established a geologic time scale that can be applied to all parts of the world.

Although the existence of fossils had been known since the early Greeks, their significance as a geologic tool was not discovered until the late 18th century. In the 1790s, the English civil engineer William Smith found that rock formations in the canals he built across Great Britain contained fossils significantly different from those in the beds above or below. He noted that layers from two different sites could be regarded as equivalent in age as long as they contained the same fossils. Therefore, sedimentary strata in widely separated areas could be identified by their distinctive fossil content. Furthermore, one type of bed such as sandstone might grade into a different bed such as limestone that contained the same fossils, indicating they were the same age.

Using the characteristics of the different strata and their fossils, Smith drew geologic maps of the varied rock formations throughout Britain. He made the most significant contribution to the understanding of fossils when he proposed the law of faunal succession, which stated that rocks could be placed in their proper time sequence by studying their fossil content. This law became the basis for the establishment of the geologic time scale and the beginning of modern geology.

The French geologists Georges Cuvier and Alexandre Brongniart refined this approach with their discovery that certain fossils in rocks around Paris were confined to specific beds. The geologists arranged fossils in a chronological order and noticed a systematic variation according to their positions in the geologic column. Fossils in the higher rock layers more closely resembled modern species than those farther down. The fossils did not occur randomly but in a determinable order from simple to complex. Units of geologic time could thus be identified by their distinctive fossil content.

In 1830, the British geologist Charles Lyell took these ideas one step further by proposing that rock formations and other geologic features took shape, eroded, and re-formed at a constant rate throughout time according to the principle of uniformitarianism—the concept that the present is the key to the past. In other words, the forces that shaped the Earth are uniform and operated in the past much as they do today. The theory was originally developed in 1785 by Lyell's mentor, the Scottish geologist James Hutton, known today as the "father of geology."

The history of the Earth has been divided into units of geologic time according to the type and abundance of fossils present in the strata. The periods take their names from the localities with the best exposures (Fig. 32). For example, the Jurassic period is named for the Jura Mountains in Switzerland, whose limestones provide a suite of fossils that adequately depicts the period.

Stratigraphic units are classified into erathems, consisting of the rocks formed during an era of geologic time. Erathems are divided into systems, consisting of rocks formed during a period of geologic time. Systems are divided into groups, consisting of rocks of two or more formations that contain common features. Formations are classified by distinctive features in the rock and are given the name of the locality where they were originally

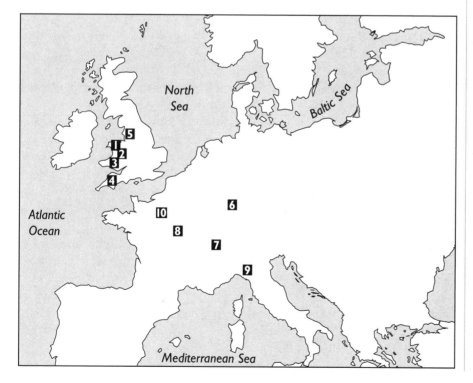

Figure 32 Type locations for geologic periods: (1) Cambrian, (2) Ordovician, (3) Silurian, (4) Devonian, (5) Carboniferous, (6) Triassic, (7) Jurassic, (8) Cretaceous, (9) Tertiary, (10) Quaternary.

described. Formations are divided into members, which might be further divided into individual beds such as sandstone, shale, or limestone.

A type section is a sequence of strata that was originally described as constituting a stratigraphic unit and serves as a standard of comparison for identifying similar widely separated units. Preferably a type section is selected in an area where both the top and bottom of the formation are exposed. Type sections are named for the area where they are best exposed. For example, the Jurassic Morrison Formation, which is well known for its dinosaur bones (Fig. 33), is named for the town of Morrison, outside Denver, Colorado.

Type sections are also distinguished by their distinct fossil content, which is used to correlate stratigraphic units. These are placed in order by age into a geologic column and are used to establish a geologic time scale. Dated material is used to place actual ages on units of geologic time. Finally, a geologic

Figure 33 *A dinosaur bone fragment in brown sandstone in the Morrison Formation, on the north side of Alameda Parkway, east of Red Rocks Park, Jefferson County, Colorado.*

(Photo by J. R. Stacy, courtesy USGS)

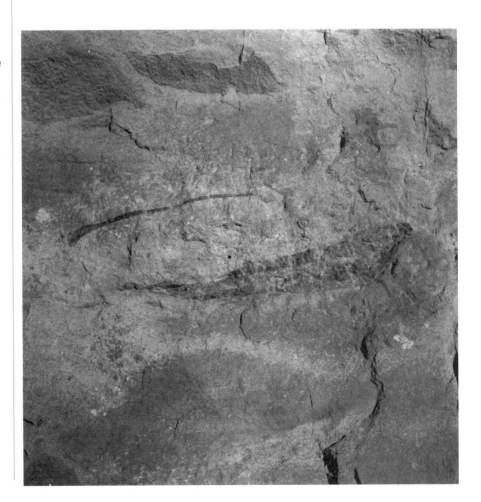

map presents in plan view the geologic history of an area where particular rock types are found.

Having considered the importance of fossils to paleontology, the next chapter will take a look at rock types that contain fossils.

3

ROCK TYPES
THE ROCK-FORMING PROCESS

Since all fossil remains and mineral finds are extracted from rocky beds, this chapter will consider first how rocks are formed, then the following chapters will discuss how fossils and minerals are deposited, shaped, located, and extracted.

A rock is a consolidated mass of earth material and can range from a diamond, the hardest known natural substance, to ice, also considered a mineral. Generally, however, rocks are siliceous materials composed of the chemical elements oxygen, silicon, aluminum, iron, calcium, sodium, magnesium, and potassium, which together account for over 90 percent of the Earth's crust. Oxygen is the most abundant element on Earth, constituting over 90 percent of the rocks, and the wide variety of minerals is determined by all the possible arrangements of oxygen atoms. Of the roughly 2,000 known minerals, only about 20 are common, and fewer than half of these constitute over 90 percent of all rocks.

The three major rock groups are igneous, sedimentary, and metamorphic (Table 5). Igneous rocks are derived directly from molten rock, or magma, which either invades the crust to form granitic rocks or erupts on the Earth's surface to produce volcanic rocks. Sedimentary rocks are cemented

TABLE 5 CLASSIFICATION OF ROCKS

Group	Characteristics	Environment
IGNEOUS		
Intrusives	Granite: mostly quartz and potassium feldspar with mica, pyroxene, and amphibole	Deep-seated, coarse-grained pluton
	Syenite: mostly potassium feldspar with mica, pyroxene, and amphibole	Deep-seated, medium-grained pluton
	Monzonite: plagioclase and potassium feldspar with mica, pyroxene, and amphibole	Deep-seated, course-grained pluton
	Diorite: mostly plagioclase and quartz with abundant mica, pyroxene, and amphibole	Deep-seated, coarse-grained pluton
	Gabbro: equal amounts of plagioclase and mica, pyroxene, and amphibole	Intermediate-depth, medium- to coarse-grained pluton
	Peridotite: mostly olivine, pyroxene, and amphibole with little plagioclase	Very-deep-seated medium- to fine-grained pluton
Extrusives	Rhyolite: mostly quartz and potassium feldspar with mica, pyroxene, and amphibole	Fine-grained fissure or volcanic eruption
	Andesite: mostly plagioclase and quartz with abundant mica, pyroxene, and amphibole	Fine-grained fissure or volcanic eruption
	Basalt: equal amounts of plagioclase and mica, pyroxene, and amphibole	Fine-grained fissure or volcanic eruption
METAMORPHIC		
Foliated	Gneiss: mostly quartz and feldspar with mica and amphibole	Coarse-grained, deep-seated
	Schist: mostly mica and platy minerals with less quartz and feldspar	Coarse-grained, deep-seated
	Phyllite: micaceous rock intermediate between schist and slate	Medium-grained, moderate depth
	Slate: feldspar, quartz, and micaceous minerals	Fine-grained, moderate depth
Nonfoliated	Hornfels: metamorphic clay material	Contact with hot magma bodies
	Marble: metamorphic carbonates	Coarse-grained, deep-seated
	Quartzite: metamorphic sandstone	Fine-grained, deep-seated
SEDIMENTARY		
Clastic	Conglomerate: fragments of rounded gravel-size sediments	River and glacial deposits
	Breccia: fragments of angular gravel-size sediments	River and volcanic deposits
	Sandstone: coarse-grained quartz and feldspar with minor accessory minerals	Marine and river deposits

continued

51

TABLE 5   CONTINUED

Group	Characteristics	Environment
	Siltstone: fine-grained quartz and feldspar with minor accessory minerals	Marine, lake, and river deposits
	Shale: very-fine-grained sediments, mostly feldspar	Marine and lake deposits
Nonclastic	Limestone: calcium carbonate, often with skeletal fragments	Marine and lake deposits
	Dolomite: calcium magnesium carbonate	Marine deposits and veins
	Gypsum: hydrous calcium sulfate	Near-shore brine pools
	Chalcedony: microscopic silica	Deep marine and groundwater

particles or grains derived from igneous, metamorphic, or other sedimentary rocks. They also precipitate directly from seawater by biologic and chemical processes. Metamorphic rocks originate as igneous or sedimentary rocks, which undergo major changes as a result of the intense temperatures and pressures deep inside the Earth. Each of these classes contains various rock specimens with specific characteristics that make them readily identifiable.

IGNEOUS ROCKS

The first rocks to form on Earth were igneous, derived directly from molten magma lying deep in the planet's interior. Most igneous rocks arise from new material in the mantle, some are derived from the subduction of oceanic crust into the mantle, and others result from the melting of continental crust. The first two types continuously build the continents, whereas the latter type adds nothing to the total volume of continental crust.

Igneous rocks are mostly silicates, which are compounds of silica and oxygen that contain metal ions. They are not simple chemical compounds, however, because their composition is not determined by a fixed ratio of atoms. Often two or more compounds are present in a solid solution with each other. In this manner, the components can be mixed in any ratio over a wide range. Most igneous rocks are aggregates of two or more minerals. Granite, for example, is composed almost entirely of quartz and feldspar with a minor constituent of other minerals. Granitic rocks formed deep within the crust, and crystal growth was controlled by the cooling rate of the magma and the available space.

Igneous rocks differ widely according to their mineral content. They generally contain auxiliary minerals such as micas, olivine, hornblende, and pyroxene. Rocks with the least iron-magnesium minerals such as amphiboles and pyroxenes are generally less dense and are light in color. They contain a high proportion of silica, usually greater than 60 percent, and are therefore called siliceous or acidic rocks. Those rich in iron and magnesium and containing a low proportion of silica, usually less than 50 percent, are generally darker with a higher density and are called basic or mafic rocks. Equivalent terms used to describe rocks that make up continental and oceanic crust are *sial* (for silica and alumina) and *sima* (for silica and magnesium).

The two major classes of igneous rocks are intrusives, derived from the invasion of the crust by a magma body from below, and extrusives, derived from the eruption of magma onto the Earth's surface from a fissure or a volcano. Both types of rocks share much the same chemical composition but have different textures because magma that pours out on the surface as lava tends to cool more rapidly, producing finer crystals. Intrusive bodies take much longer to cool because the rocks they invade make good insulators and tend to hold in the heat. This slow cooling rate allows large crystals to grow, and generally the larger the magma body the longer molten rock takes to cool and consequently the larger the crystals.

Intrusive magma bodies have several shapes and sizes (Fig. 34). The largest are batholiths, which have more than 40 square miles of surface exposure and are usually longer than they are wide. Batholiths produce some of the major mountain ranges, such as the Sierra Nevada in California (Fig. 35), which are nearly 400 miles long and about 50 miles wide. Batholiths are composed of granitic rocks with large crystals, mostly quartz, feldspar, and mica. Granites found at the center of a batholith tend to be more coarsely grained than those at the edges, because the center cools more slowly than the outside, allowing larger crystals to form. The rocks might contain regions where ores have accumulated in veins, which were formed when metal-rich fluids from a magma chamber migrated into cracks and fractures in the rocks. For this reason, these mountains are favorite hunting grounds for prospectors in search of gold, silver, and other valuable minerals.

An intrusive magma body, shaped much like a batholith but with less than 40 square miles of surface exposure, is called a stock. It is generally circular or elliptical in form and might be a projection of a larger batholith buried deeper down. Stocks also are composed of coarse-grained granitic rocks. When exposed by erosion, stocks can form lone mountains, sometimes standing in the middle of nowhere.

If an intrusive magma body is tabular in shape and is considerably longer (on the order of several miles) than it is wide (only a few feet), it indicates that the magma fluids occupied a large crack or fissure in the crust, thus forming a dike. Dikes are often outgrowths from a batholith that normally cut across existing rock structures. Because dike rocks are usually harder than the sur-

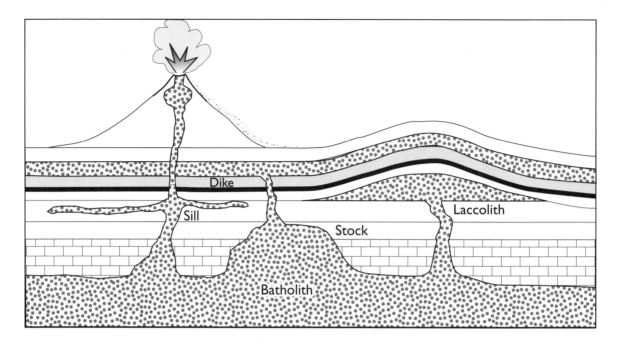

Figure 34 *A cutaway view of intrusive magma bodies that invade the Earth's crust and erupt on the surface as volcanoes.*

Figure 35 *The Sierra Nevada Range, Inyo County, California.*

(Photo by W. C. Mendenhall, courtesy USGS)

rounding material, they generally form long ridges when exposed by erosion. Large dikes can be seen radiating outward from Shiprock (Fig. 36), which is a 1,300-foot volcanic neck located in the northwestern corner of New Mexico.

Sills are similar to dikes in their tabular form but are produced parallel to planes of weakness such as sedimentary beds. When a sheetlike mass of fluid magma flows between layers of rock, it lifts them to make room for more magma. The magma cools more rapidly in contact with the surrounding rock; therefore the outer zones differ greatly in structure and composition from the center of the sill. Both dikes and sills are small bodies compared to batholiths and therefore cool more rapidly, providing somewhat finer-grained granitic rocks.

Figure 36 *Shiprock, showing large dike, San Juan County, New Mexico.*

(Photo by W. T. Lee, courtesy USGS)

A special type of sill is called a laccolith; it tends to bulge the overlying sediments upward, sometimes forming mountains such as the Henry Mountains in southern Utah. When magma spreads outward between sedimentary layers, it forces them upward, often to a height of 1,000 feet or more over an area of 100 square miles. The formation has a definite floor that is revealed when erosion cuts through rock layers beneath the intrusion.

A volcanic neck or plug results when magma fills the throat of a volcano without erupting and cools slowly to form rocks similar to those of dikes. It is a vertical cylindrical body, ranging up to one mile in diameter, composed of solidified magma that once filled the main conduit or pipe of a volcano; erosion leaves the more resistant rock standing above the surrounding terrain. The most prominent volcanic neck in North America is Devils Tower in northeastern Wyoming (Fig. 37).

Igneous rocks are classified according to their mineral content and texture, which in turn are governed by the degree of separation and rate of cooling of the magma. The first mineral to form with falling temperatures is olivine (Fig. 38), followed by pyroxene, amphibole, and biotite for the iron-magnesium silicates, or sima. For the aluminum silicates, or sial, calcium and sodium feldspar, called plagioclase, are the first minerals to form. Upon further cooling of the magma, sima and sial grade into potassium feldspar (orthoclase and microcline), followed by muscovite and finally quartz, the mineral that forms at the lowest-temperature. Texture is controlled by the rate of cooling, as the slowest rate gives rise to the largest crystals and the more rapid rates provide smaller crystals, until the cooling becomes so rapid that a natural glass, called obsidian, forms.

The following are the major igneous rocks:

1. Granite, consisting mostly of coarse-grained quartz and potassium (pink) feldspar

Figure 37 *Devils Tower, Cook County, Wyoming.*

(Courtesy USGS)

2. Syenite, which is similar in texture to granite but with little or no quartz
3. Monzonite, which has about equal proportions of plagioclase (white) feldspar and potassium feldspar with abundant dark accessary minerals
4. Diorite, which resembles monzonite but with plagioclase the dominant feldspar
5. Gabbro, whose chief minerals are pyroxene and plagioclase feldspar
6. Peridotite, consisting principally of olivine and pyroxene
7. Rhyolite, the volcanic equivalent of granite

8. Andesite, the volcanic equivalent of diorite
9. Basalt, the volcanic equivalent of gabbro and the most abundant of all lavas

If magma extrudes onto the Earth's surface either through a fissure eruption, the most prevalent kind, or a volcanic eruption, which builds majestic mountains (Fig. 39), it produces a variety of rock types depending on the source material, which in turn controls the type of eruption. Ejecta from volcanoes have a wide range of chemical, mineral, and physical properties (Table 6). Nearly all volcanic products are silicate rocks, composed mainly of oxygen, silicon, and aluminum, with lesser amounts of iron, calcium, magnesium, sodium, and potassium. Basalts are relatively low in silica and high in calcium, magnesium, and iron. Magmas that have larger amounts of silica, sodium, and potassium along with lesser amounts of magnesium and iron form rhyolites, which contain mostly quartz crystals, and andesites, which contain mostly feldspar crystals.

All solid particles ejected into the air by volcanic eruptions are collectively called tephra, from the Greek, meaning "ash." (Actually, the word *ash* is a historical misnomer left over from the days when volcanoes were thought to arise from the burning of subterranean substances.) Tephra includes an assortment of fragments from dust-size material to large blocks. It results when

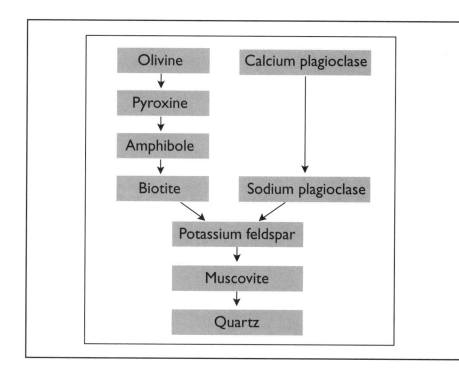

Figure 38 Sequence of mineral formation with falling temperature.

Figure 39 *Eruption of*
Mount Saint Helens on
May 18, 1980.

(Courtesy USGS)

Figure 39 *Eruption of Mount Saint Helens on May 18, 1980.*

(Courtesy USGS)

molten rock containing volatiles, which help make magma flow easily and are composed of water and dissolved gases, mostly carbon dioxide, rises through a conduit and suddenly separates into liquid and bubbles as it nears the surface. With decreasing pressure, the bubbles grow larger. If this event occurs near the orifice, a mass of froth might spill out and flow down the sides of the volcano,

TABLE 6   CLASSIFICATION OF VOLCANIC ROCKS

Property	Basalt	Andesite	Rhyolite
Silica content	Lowest about 50%, a basic rock	Intermediate about 60%	Highest more than 65%, an acid rock
Dark mineral content	Highest	Intermediate	Lowest
Typical minerals	Feldspar Pyroxene Olivine Oxides	Feldspar Amphibole Pyroxene Mica	Feldspar Quartz Mica Amphibole
Density	Highest	Intermediate	Lowest
Melting point	Highest	Intermediate	Lowest
Molten rock viscosity at the surface	Lowest	Intermediate	Highest
Formation of lavas	Highest	Intermediate	Lowest
Formation of pyroclastics	Lowest	Intermediate	Highest

forming pumice, which, because of trapped air inside glass vesicles, can actually float on water. A classic example of this effect occurred during the 1883 eruption of the Indonesian volcano Krakatoa, which sent huge blocks of floating pumice into the sea and threatened shipping in the area.

If the reaction occurs deep in the throat of a volcano, the bubbles might expand explosively and burst the surrounding liquid, fracturing the magma into fragments. Like pellets from a shotgun, the fragments are driven upward by the force of the expansion and hurled high above the volcano. The fragments cool and solidify during their flight through the air. Blobs of still-fluid magma called volcanic bombs might splatter the ground nearby. If they cool in flight, they form a variety of shapes, depending on how fast they are spinning, which can make them whistle as they fly through the air. If the bombs are about the size of a nut, they are called *lapilli,* Latin for "little stones," and form strange gravellike deposits along the base of a volcano.

Tephra supported by hot gases produced by a lateral blast of volcanic material is called *nuée ardente,* French for "glowing cloud." The cloud of ash and pyroclastics flows streamlike near the ground and might follow existing river valleys for tens of miles at speeds upward of 100 miles per hour. The best-known example was the 1902 eruption of Mount Pelée, Martinique, which in minutes annihilated 30,000 inhabitants. When the tephra cools and solidifies, it forms deposits called ash-flow tuffs that can cover an area as much as 1,000 square miles or more.

Welded tuffs are consolidated pumice, lapilli, and glass shards in an ash flow. The welded volcanics and weight of the overlying materials collapse bubbles and flatten elongated fragments to produce laminations in the tuff, which give it a banded appearance. *Ignimbrites,* from Greek meaning "raining fire," are formed by solidified pyroclastic material deposited by nuée ardentes and ash flows. Some of the largest ignimbrite sheets in the world lie in the Altiplano region high in the Andes Mountains of South America.

Nearly all volcanoes produce some tephra. Even relatively quiet eruptions occasionally eject fountains of highly fluid lava, whose spray solidifies into a minor amount of tephra that is confined to the neighborhood of the vent. If water from the sea, a lake, or a water table enters the magma chamber, it instantaneously flashes into steam, causing violent explosions to rise through the conduit accompanied by little or no new magma. Most tephra produced in this manner originates from the walls of the conduit or from shattered parts of the crater.

Lava is molten magma that reaches the throat of a volcano or the top of a fissure vent without exploding into fragments and is able to flow onto the surface. The magma that produces lava is much less viscous, or more fluid, than the magma that produces tephra. This allows volatiles and gases to escape more easily and gives rise to much quieter and milder eruptions. Lava is mostly composed of basalt, which contains only about 50 percent silica, and is dark–colored and quite fluid. The outpourings of lava have two general classes that take Hawaiian names and are typical of Hawaiian eruptions: *pahoehoe* (pronounced pah-HOE-ay-hoe-ay), which means "satinlike," and *aa* (pronounced AH-ah), inspired by the sound of the pain that results from walking over them barefooted.

Pahoehoe, or ropy lavas (Fig. 40), are highly fluid basalt flows produced when the surface of the flow congeals and forms a thin plastic skin. The melt beneath continues to flow outward, molding and remolding the skin into bil-

Figure 40 *Pahoehoe lava flow from an eruption of Kilauea Volcano, Hawaii.*

(Photo by D. A. Swanson, courtesy USGS)

lowing or ropy-looking surfaces. When the lava eventually solidifies, the skin retains the appearance of the flow pressures exerted on it from below. Aa, or blocky lava, forms when viscous, subfluid lava presses forward, carrying a thick, brittle crust along with it. As the lava flows, it stresses the overriding crust, breaking it into rough, jagged blocks, which are pushed ahead of or dragged along with the flow in a disorganized mass.

Highly fluid lava moves rapidly, especially down the steep slopes of a volcano. The speed of the flow is also determined by the viscosity of the lava and the time it takes to harden. Most lavas flow at a walking pace to about 10 miles per hour. Some lava flows have been clocked at only a snail's pace; others move as fast as 50 miles per hour. Some very thick lavas creep ahead slowly for months or even years before they finally solidify. Sometimes, the upper layers of a basalt lava flow contain gas bubbles that when cooled show numerous irregular smooth-walled holes, producing a dark-colored rock called scoria. A certain type of scoria called a volcanic cinder resembles clinkers or cinders from a coal furnace.

After a stream of lava has crusted over and hardened on the surface, if the underlying magma continues to flow away, a long lava tube or lava cave is formed. Lava tubes can reach 30 feet or more across and extend for hundreds of feet. The walls and roof of a lava cave are occasionally adorned with stalactites, and the floor is covered with stalagmites composed of deposits of lava. In some cases, especially on the ocean floor, the lava solidifies into tubular-shaped masses called pillow lava. As a lava flow cools, it shrinks, causing cracking or jointing. The cracks can shoot vertically through the entire lava flow, breaking it into six-sided pillars or columns like those found at Devils Postpile National Monument in east central California. (Fig. 41).

SEDIMENTARY ROCKS

Sedimentary rocks are derived from the weathering or the decomposition and disintegration of older rocks, including igneous, metamorphic, and other sedimentary rocks. The two basic classes of sedimentary rocks are clastics, comprising particles or grains, and precipitates, comprising water-dissolved minerals, mostly calcium and silica. Clastic sedimentary rocks are composed mainly of fragments broken loose from a parent material, deposited by mechanical transport, and cemented into hard rock. In addition, dissolved minerals in groundwater can cement clastic particles together to form solid rock beneath the Earth's surface.

Rocks are weathered, or broken down, into sediment grains by the action of water and wind, cycles of heating and freezing, and the activities of plants and animals. Weathering causes rocks to break apart or causes the outer layers to peal or spall off in a process known as exfoliation. The products of weathering include a range of materials from very fine-grained sediments to large boulders. Erosion by water, wind, rain, or glacial ice eventually takes the sediments to streams and rivers, which in turn empty into the ocean. The more angular the sediment grains, the less time they have spent in transit, whereas the more rounded sediment grains indicate severe abrasion caused by travel across long distances or reworking by fast-flowing streams or pounding waves.

When the suspended sediments reach the ocean, they settle out under the influence of gravity according to grain size, with the coarser-grained sediments settling out near the turbulent shore and the finer-grained sediments settling out in calmer waters farther out to sea. As the shoreline advances toward the sea, as a result of the buildup of coastal sediments or falling sea levels, finer sediments are progressively covered over by coarser ones. As the shoreline recedes as a result of rising sea levels, coarser sediments are covered by progressively finer ones. This process provides a recurring sequence of sandstones, siltstones, and shales (Fig. 42).

Terrestrial sediments are formed entirely on land and include deposits of windblown sand (called eolian deposits) that are identified by sand dunes; river deposits, identified by cross-bedding, and ripple marks; lake or marsh deposits (that can contain fossils and coal beds); and glacial deposits, distinctively heaped sediments deposited in places where the glaciers melted. Sand moves across the desert floor by a process known as saltation (Fig. 43). When a windblown sand grain lands, its momentum kicks up another sand grain, which,

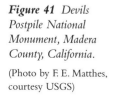

Figure 41 *Devils Postpile National Monument, Madera County, California.*

(Photo by F. E. Matthes, courtesy USGS)

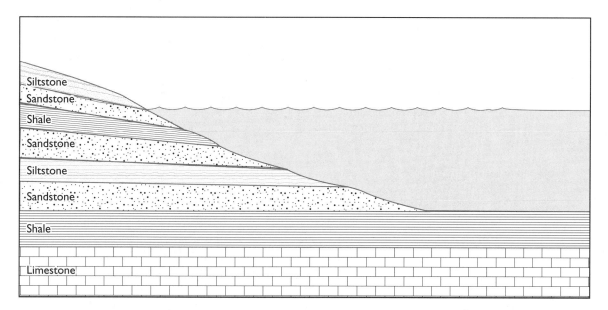

Siltstone
Sandstone
Shale
Sandstone
Siltstone
Sandstone
Shale
Limestone

when it falls, kicks up another in a sort of chain reaction. This process forms sand dunes that march across the desert floor in the direction of the prevailing wind, engulfing everything in their path. The process can be particularly troublesome in regions bordering deserts such as the Sahel region south of the Sahara Desert in Africa. Desertification is becoming a serious problem in many parts of the world, turning once-fertile agricultural land into desert by the destructive activities of people and nature.

Clastic sedimentary rocks are generally classified according to grain size. Gravel-size sediments are called conglomerates if rounded and breccia if angular. They are composed of abundant quartz and chalcedony (microcrystalline quartz, that is, flint or chert). Breccias are relatively rare and are indicative of terrestrial mudflows or submarine slides and earthquake faults. Debris flows piled up on the continental slope can produce a coarse carbonate rubble known as brecciola. Volcanic breccias, also known as agglomerates, are consolidated pyroclastic fragments. Glacial deposits composed of boulders and gravel-size sediments are called tillites.

Sandstones are composed mostly of quartz grains roughly the size of beach sands. Indeed, many sandstones such as the St. Peter Sandstone of the central United States, which is used for the manufacture of glass, were once beach deposits. If a sandstone has abundant feldspar as well as quartz, it is called an arkose. Graywacke, sometimes called dirty sandstone, is a dark, coarse-grained sandstone with a clay matrix and is believed to be deposited by submarine turbidity currents. Siltstones are composed of fine quartz grains that are just visible to the naked eye. Shales or mudstones are composed of the

Figure 42 *A stratigraphic cross section showing a sequence of sandstones, siltstones, and shales overlying abasement rock composed of limestone.*

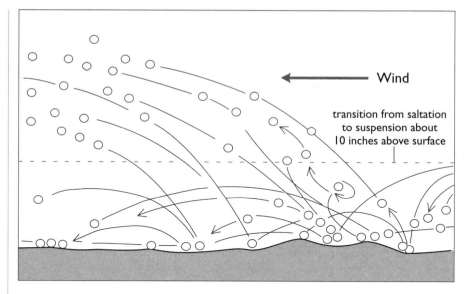

Figure 43 *The process of saltation causes sand to march across the desert floor.*

Wind

transition from saltation
to suspension about
10 inches above surface

finest sedimentary particles in clay or mud, whose grains are invisible to the unaided eye.

Shales and mudstones are the most abundant sedimentary rocks because they are the main weathering products of feldspars, which are the most abundant minerals. Furthermore, all rocks are eventually ground down to clay-size particles by abrasion. Because clay particles are small and sink slowly, they normally settle out in calm, deep waters far from shore. Compaction, resulting from the weight of the overlying sediments, squeezes out water between sediment grains, and the clay is lithified into mudstone if massive or shale if fissile, or thinly bedded. Clastic sediments that comprise particles or grains are lithified into rock mainly by compaction for fine-grained sediments and cementation for coarse-grained, well-sorted sediments, which have grains of equal size. As increasing layers of sediments pile up, water is squeezed out of the lower strata by the overlying weight until individual grains are pressed together. Older sediments might show a higher degree of compaction because they were buried more deeply, although the degree of compaction is not always a reliable indicator of age.

Minerals such as calcium carbonate and silica are dissolved in groundwater, which flows around coarse sediment grains. The minerals are deposited between sediment grains and cement them together. If iron oxides are the cementing agents, they tend to color the rocks red, brown, or yellow. Such colors typify terrestrial sediments. If clays are used as cementing agents for poorly sorted marine sediments, which have grains of varying sizes, they tend to color the rock gray or gray-green.

Nonclastic, or precipitate, rocks such as limestones are formed by biologic and chemical precipitation of minerals dissolved in water. Rainwater normally has a small amount of carbonic acid (the same acid in soft drinks) from the chemical reaction of water and carbon dioxide in the atmosphere. This acid plays an important role in dissolving calcium and silica minerals from surface rocks to form bicarbonates. The bicarbonates enter rivers that reach the ocean and become thoroughly mixed with seawater by the action of waves and currents.

The bicarbonates precipitate mostly by biologic activity as well as direct chemical processes. Living organisms require calcium bicarbonate to build supporting structures such as shells composed of calcium carbonate. When the organism dies, its skeleton falls to the ocean floor, where over time thick deposits of calcium carbonate called calcite ooze build up to form limestone (Fig. 44). Limestone, the most common precipitate rock, is mostly produced by biologic activity. This is evidenced by the abundance of fossils of marine life in limestone beds.

If limestone is composed almost entirely of fossils or their fragments, it is called a coquina. Some limestone is chemically precipitated directly from seawater, and a minor amount precipitates in evaporite deposits from brines. Most limestones originated in the ocean, and some thin limestone beds were deposited in lakes and swamps. Many limestones form massive formations (Fig. 45), which are recognized by their typically light gray or light brown color. Whole or partial fossils constitute many limestones, depending on whether they were deposited in quiet or agitated waters.

Figure 44 *Formation of carbonate sediment on the ocean floor from the burial of skeletons of marine organisms.*

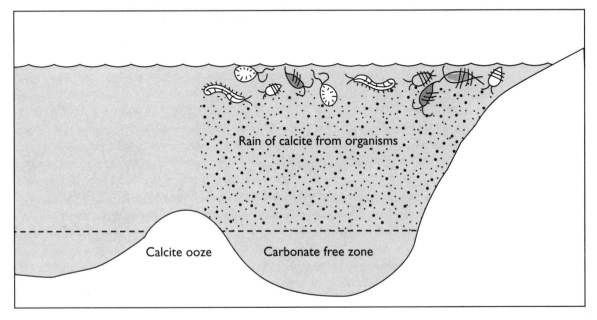

Rain of calcite from organisms

Calcite ooze   Carbonate free zone

Figure 45 *Intensely fold-ed limestone formation in Atacama Province, Chile.*

(Photo by K. Segerstrom, courtesy USGS)

Chalk is a soft, porous carbonate rock and should not be confused with the chalk used on classroom blackboards, which is actually composed of calcium sulfate. Thick beds of chalk, which were deposited during the Cretaceous, gave the period its name—*creta* is Latin for "chalk." Among the largest chalk deposits are the chalk cliffs of Dorset, England, which, because they are so soft, severely erode during violent coastal storms.

Dolomite, which resembles limestone, is produced by the partial replacement of calcium in limestone with magnesium. The replacement can cause a reduction in volume, forming void spaces, which can destroy any fossils that are present. For the last two centuries, geologists have been puzzled by the so-called dolomite problem. Ancient dolomite deposits were laid down in huge heaps such as those that *created* the Dolomite Alps in northern Italy, but today little dolomite is being formed. Dolomite appears to have been made from the excrement of sulfate-consuming bacteria that apparently were far more prevalent in the past.

The majority of carbonate sediments were deposited in shallow seas, generally less than 100 feet deep, and mainly in intertidal zones, where marine organisms were plentiful. Coral reefs, which form in shallow water where sunlight can easily penetrate to allow photosynthesis, contain abundant organic remains. Many ancient carbonate reefs are composed largely of carbonate mud that contains larger skeletal remains.

Most carbonate rocks began as sandy or muddy calcium carbonate material. The sand-size particles are composed of broken-up skeletal remains of invertebrates and shells of calcareous algae that rain down from above. The skeletal remains might have been broken up by mechanical means, such as the pounding of the surf, or by the activity of living organisms. Further breakdown into dust-size particles produces a carbonate mud, which is the most common constituent of carbonate rocks, forming a matrix known as micrite. Under certain conditions, the carbonate mud dissolves in seawater and is redeposited elsewhere on the ocean floor, forming a calcite ooze that is later lithified into limestone.

As calcareous sediments accumulate into thick deposits on the ocean floor, deep burial of the lower strata produces high pressure, which lithifies the beds into carbonate rock, consisting mostly of limestone or dolomite. If fine-grained calcareous sediments are not strongly lithified, they form deposits of soft, porous chalk. Limestones typically develop a secondary crystalline texture, resulting from the growth of calcite crystals by solution and recrystallization after the formation of the original rock.

Some carbonate rocks were deposited in deep seas. The maximum depth at which carbonate rocks can form is determined by the calcium carbonate compensation zone, which generally begins at a depth of about two miles. Below this zone, the cold, high-pressure waters of the abyss, which contain the vast majority of free carbon dioxide, dissolve calcium carbonate sinking to this level. The upwelling of deep ocean water, mainly in the Tropics, returns to the atmosphere carbon dioxide lost by the carbon cycle, which is the circulation of carbon by geochemical processes (Fig. 46). Silica readily dissolves in seawater in volcanically active areas on the seafloor and as a result of volcanic eruptions into the sea and weathering of siliceous rocks on the continents. Some organisms such as diatoms (Fig. 47) extract the dissolved silica directly from seawater to build their shells and skeletons. Accumulations of siliceous sediment on the ocean floor from dead organisms form diatomaceous earth, also called diatomite. Thick deposits throughout the world are a tribute to the prodigious growth of these organisms during the last 600 million years. If diatomite is lithified into solid rock, it forms chalcedony, which also forms by direct precipitation from seawater. Chalcedony produces varieties of opal (which can recrystallize into chert), banded agate, jasper, and flint.

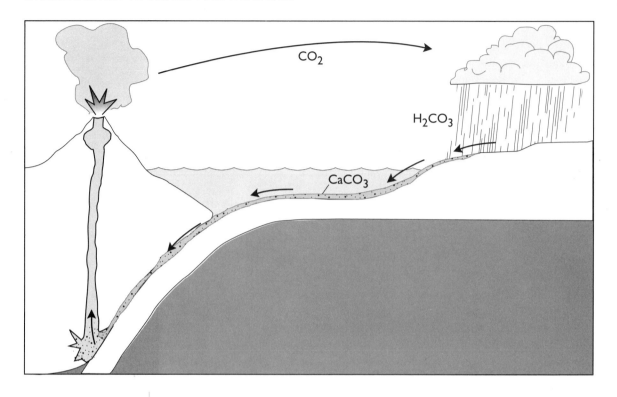

CO_2

H_2CO_3

$CaCO_3$

Figure 46 *The carbon cycle. Carbon dioxide converted into bicarbonate is washed off the land and enters the ocean, where marine organisms convert it into carbonate sediments, which are thrust into the Earth's interior and become part of the molten magma. Carbon dioxide returns to the atmosphere by volcanic eruptions.*

Evaporite deposits are produced in arid regions near shore, where pools of brine, which are constantly replenished with seawater, evaporate in the hot sun, leaving salts behind. Evaporite deposits generally form under arid conditions in areas that lie between 30 degrees north and south of the equator. However, extensive salt deposits are not being formed at present, suggesting a cooler current global climate. The fact that ancient evaporite deposits exist as far north as the Arctic regions indicates either that these areas were at one time closer to the equator or the global climate was considerably warmer in the geologic past. Evaporite accumulation peaked about 230 million years ago, when the supercontinent Pangaea was beginning to rift apart. Few evaporite deposits date beyond 800 million years ago, however, probably because most of the salt formed before then has been recycled back into the ocean.

The salts precipitate out of solution in stages. The first mineral to precipitate is calcite, closely followed by dolomite, although only a small amount of limestone and dolostone is produced in this manner. After about two-thirds of the water is evaporated, gypsum precipitates. When nine-tenths of the water is removed, halite, or common salt, forms. Thick deposits of halite are also produced by the direct precipitation of seawater in deep basins that have been cut off from the general circulation of the ocean such as the Mediterranean and the Red seas.

Gypsum, present in thick beds composed of hydrous calcium sulfate, is one of the most common sedimentary rocks. The beds are produced in evaporite deposits that form when a pinched-off portion of the ocean or an inland sea evaporated. Oklahoma, like many parts of the interior of North America that were invaded by a Mesozoic sea, is well known for its gypsum beds. The mineral is mined extensively for the manufacture of plaster.

Coal is generally regarded as a sedimentary rock, even though it did not originate from clastic sediments or result from chemical precipitation. Coal

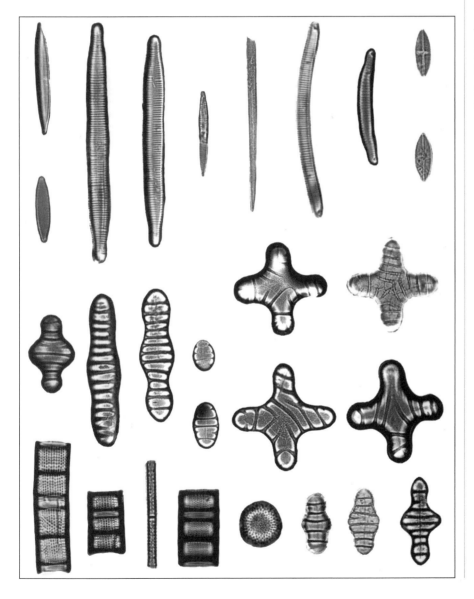

Figure 47 Late Miocene marine diatoms from the Kilgore area, Cherry County, Nebraska.

(Photo by G. W. Andrews, courtesy USGS)

originated compacted plant life that grew in lush swamps. Often between easily separated layers of coal or associated fine-grained sedimentary beds are carbonized remains of ancient plant stems and leaves. Black or carbonized shales also originated in the ancient coal swamps, and traces of plant life can be found between shale layers.

METAMORPHIC ROCKS

Igneous and sedimentary rocks that have been subjected to intense temperatures and pressures of the Earth's interior, heat generated near magma bodies, shear pressures from earth movements, or strong chemical reactions that do not cause the rocks to melt are called metamorphic rocks, so named because of the material changes that have taken place within the rocks themselves. Chemical changes between existing minerals might produce large crystals of garnet or other common metamorphic minerals. Metamorphism causes dramatic changes in texture, mineral composition, or both, often making determination of the nature of the original rock difficult.

The most noticeable effect of metamorphism on a rock is foliation, which is closely spaced layering due to the parallel orientation of some minerals. During metamorphism, the increased pressure squeezes and mashes grains as they form, causing them to foliate. Minerals whose crystals form thin plates are most likely to form foliated patterns. The minerals are often drawn out, flattened, and arranged in parallel layers or bands. Rocks that contain mica and iron-magnesium minerals show strong foliation because they tend to form flakes or needles that grow larger in only one direction. Rocks containing quartz or feldspar do not become foliated because under pressure they tend to grow in all directions. Foliated rocks often split easily parallel to the banding.

Metamorphism produces new textures by recrystallization, whereby minerals grow into larger crystals, which might have a different orientation. The crystals grow by laying down layers upon layers of atoms (see Chapter 7 on crystals). New minerals are also created by recombining chemical elements to form new associations. Water and gases from nearby magma bodies also aid in the chemical changes taking place in rocks by conveying chemical elements from one place to another.

Although many minerals are strictly metamorphic in origin, most metamorphic rocks are similar in composition to their parent rocks. Heat is probably the most important agent for recrystallization, and often deep burial is required to generate the temperatures and pressures required for extensive metamorphism. Varying degrees of metamorphism are also achieved at shallower depths in geologically active areas with higher thermal gradients, where the temperature increases more rapidly with depth than normal.

During metamorphism, rocks behave plastically and are able to bend or stretch as a result of the high temperatures and great pressures applied by the overlying rocks. The stress might deform the rocks and flatten or stretch pebbles or fossils. Generally, most fossils are destroyed during intense or prolonged metamorphism as the rocks recrystallize under stress, erasing all evidence of fossils.

Because any rock type can be metamorphosed, a wide range of metamorphic rocks exist, but basically they can be classified into two major categories: foliated with a layered or banded structure and nonfoliated with a massive structure. Mountain building, which provides the forces necessary for folding and faulting rocks at shallow depths, also provides the stress forces needed to produce foliated metamorphic rocks deeper down. The most common foliated metamorphosed rocks are called schist and gneiss. When a rock undergoes successive periods of metamorphism, one type of metamorphic rock might change into another. For example, slate usually forms schist, and schist generally forms gneiss.

A schist is an intensely foliated crystalline rock with much mica and little feldspar, which causes it to split readily along layers. It is a coarse-grained metamorphic rock formed from slate under high pressure. Schists are strongly foliated rocks with fairly distinct bands of different minerals and usually split very easily along the bands. They are classified according to the most prominent mineral in the rock, such as mica, hornblende, chlorite, or quartz. A spectacular crystalline schist called the Vishnu Schist lies on the bottom of the western end of the Grand Canyon (Fig. 48).

A gneiss is a metamorphic equivalent of granite with abundant feldspar. It is the most coarsely textured banded metamorphic rock, consisting of alternating bands. The lighter bands are rich in quartz and feldspar, and the darker bands are rich in biotite mica, hornblende, or garnet. Gneiss is derived from either granite, complex rocks of igneous or sedimentary origin, or a mixture of metamorphic rocks that have been invaded by igneous materials. Like schist, gneiss is classified by its most conspicuous mineral, but does not split nearly as well.

Low-grade metamorphism transforms shale into slate, a uniformly fine-grained rock that splits easily into smooth slabs, a property that makes it useful for such products as chalk boards and building materials. The metamorphism causes a mechanical reorientation of clay particles as well as minor recrystallization. Slate is often colored black from carbon that has been metamorphosed into graphite. Iron minerals give slate a red color and magnesium minerals give it a green color.

Intermediate in texture between a schist and slate is phyllite (Fig. 49), which tends to break into slabs with a slightly wrinkled surface. Phyllites are formed from shale under greater pressure than that which forms slate. They

Figure 48 *Precambrian Vishnu Schist, Grand Canyon National Park, Arizona.*

(Photo by R. M. Turner, courtesy USGS)

generally have a higher luster than slate and often show a silky luster due to the presence of fine grains of mica.

Nonfoliated rocks consist of hornfels, formed by contact metamorphism in narrow belts around intrusive magma bodies (Fig. 50). Their formation is similar to the baking of clay in a kiln to make pottery. This produces fine-grained, hard rocks that can be either completely recrystallized with none of the original features preserved or only slightly modified with most of the original features kept intact.

Single mineral crystalline metamorphic rocks are produced from limestone or sandstone. Marble is the product of the metamorphism of limestone or dolomite but is distinguishable from limestone by its larger crystals and sometimes spectacular fossils on a polished surface. Under heat and pressure, the fine crystalline particles of limestone are recrystallized, forming much larger crystals. Organic materials that sometimes darken the original limestone are driven off, resulting in nearly pure white marble that is valued by sculptors. Mineral impurities might show multicolored veins, bands, and other patterns in marble that is used as ornamental stone.

Quartzite is produced by the metamorphism of quartz sandstone or quartz grains cemented by silica. Its sugary luster distinguishes quartzite from

Figure 49 *A thrust fault in phyllite in cut on irrigation ditch along the east bank of the Riberao de Mata, northeast of Pires, Minas Gerais, Brazil.*

(Photo by P. W. Guild, courtesy USGS)

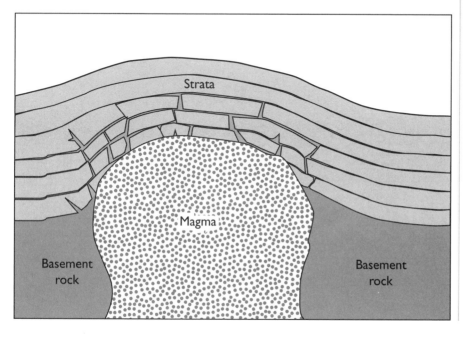

Strata

Magma

Basement rock

Basement rock

Figure 50 *Contact metamorphism is produced by the intrusion of a magma body into overlying rocks, as shown here.*

chalcedony, which has a nearly waxlike luster. Compaction, recrystallization, or cementation with silica closes the pores of the original sandstone, making quartzite extremely hard and durable. Because quartzite is highly resistant to weathering, it often forms hills and mountains after weaker rocks have been eroded away.

A special kind of metamorphism results when rocks are broken, sheared, shocked, or ground on or near the Earth's surface, where temperatures and pressures are too low to cause significant recrystallization. When a fault slips, it produces a tremendous amount of pressure and frictional heat at the point of contact between fault planes in a very short time. A metamorphic rock similar to a schist results from grinding or milling by the fault movement.

Meteorite impact craters (Fig. 51) are surrounded by circular disturbed areas, consisting of rocks that have been reworked by shock metamorphism, which changes their composition and crystal structure. Shock metamorphism requires the instantaneous application of high temperatures and pressures such as those found deep in the Earth's interior. Shocked metamorphic minerals spread around the world are believed to have originated from a massive impact that took place 65 million years ago, when the dinosaurs became extinct.

After becoming familiar with the different rock types, the next chapter provides information on the formation of fossils in the sedimentary environment.

Figure 51 *The structure of a large meteorite crater.*

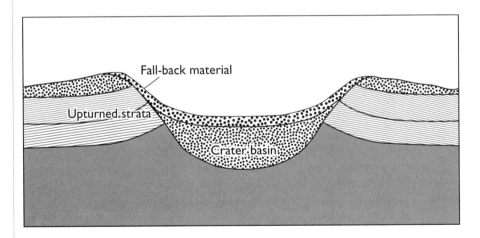

4

FOSSIL FORMATION
THE PRESERVATION OF PAST LIFE

Most of the Earth's surface is covered by a thin veneer of sediment, which preserves the vestiges of past life. Under special geologic conditions, with minimal predation or decomposition and rapid burial, the bodies of dead organisms are preserved to withstand the rigors of time. Most fossils comprise ancient marine organisms because they were the earliest form of animal life and the most abundant and therefore stood a better chance of becoming fossilized. The ocean, where sedimentation occurs, also provides the ideal setting for the preservation of species.

However, some organisms, especially those with soft bodies, are not as well preserved in the fossil record. Thus, fossils are dominated by organisms with hard skeletal remains, and shells, bones, teeth, and wood dominate the record of past life. Unfortunately, because preservation requires these rigid limitations, the fossil record is well represented by organisms with hard body parts but poorly represented, or not represented at all, by organisms with soft body parts, so that the fossil record gives a somewhat lopsided view of previous life on Earth.

THE FOSSIL FAMILY TREE

Fossils of extinct organisms are classified by the same system used for classifying living organisms. The first classification scheme was developed by the 18th-century Swedish botanist Carl von Linné, better known by his latinized name Carolus Linnaeus. He gave Latin names to organisms because Latin was the universal language of science during his time. His naming was based on the number of characteristics organisms had in common.

Linnaeus realized that some organisms had a greater similarity than others because they were more closely related. Later, as evolution was recognized as the process by which organisms develop into new species, classification schemes were developed to describe these evolutionary patterns, demonstrating how groups of organisms were related in both space and time.

In the classification scheme used by biologists and paleontologists, each organism is assigned an italicized two-part species name. The first word, which is capitalized, is the generic name and is shared with other very closely related species. The second word, written in lowercase, is the species name and is unique to a particular genus: for example, *Homo sapiens,* which is who we are.

Often the name of the discoverer of a new species and the date of discovery will follow the species name. The scientific name of a species is Greek or Latin to provide enough names for the nearly 2 million known species. Every species must have a unique name, and that name cannot be used again for another species; neither can more than one name be applied to the same species.

The classification scheme establishes a hierarchy (Table 7) in which each step up the ladder becomes more inclusive, encompassing a larger number of organisms. A kingdom comprises all the species of animals, plants, or archea (primitive bacteria). The kingdom we belong to is Animalae; our phylum is Chordata; our subphylum is Vertebrata; our class is Mammalia; our order is Primate; and our family is Hominidea. Our genus *Homo* encompasses all our ancient ancestors (Fig. 52), starting with *Homo habilis,* who lived around 2 million years ago.

TABLE 7   CLASSIFICATION OF ORGANISMS

Kingdom (plants, animals, archea)

 Phylum (33 phyla)

  Order

   Family

    Genus (average 60 species per genus)

     Species (1.75 million known species)

      Breed (groups of closely related organisms)

A taxonomical system called cladistics reformed classification by using only the branching order of lineages on evolutionary trees. Therefore, in the cladistic system, lungfish are more closely related to land vertebrates than to bony fish, even though they are more similar to the latter in outward appear-

Figure 52
Australopithecus *scavenging on the African plain.*

(Photo courtesy National Museums of Canada)

ance. This is because the common ancestor of lungfish and terrestrial vertebrates appears more recently in time than that linking the lungfish with modern bony fish. In other words, the cladistic system of classification works only with the closeness of common ancestry in time without regard to similarities in morphological characteristics or appearance.

Most fossils found are those of marine animals. More remains of marine organisms have been fossilized because seawater is a good preservative and sedimentation takes place more readily in water. In addition, marine life has been in existence about eight times longer than terrestrial life; therefore, a greater number of organisms were available for fossilization.

The vast majority of fossils are included in just 10 phyla (Table 8). The first phylum, Protozoa, begins with the simplest life forms, and each succeeding phylum—Porifera, Coelenterata, Bryozoa, Brachiopoda, Molluska, Annelida, Arthropoda, and Echinodermata—becomes more complex; the phylum Chordata, where we belong, is the most complex of all. Phyla are ordered in this manner in recognition of the evolutionary advancement of species.

The first organisms to develop on Earth, some 3.5 billion years ago, were probably bacteria and ancestral blue-green algae. These were possibly fossilized

TABLE 8   CLASSIFICATION OF FOSSILS

Group	Characteristics	Geologic Age
Protozoans	Single-celled animals: foraminifera and radiolarians, about 80,000 living species	Precambrian to recent
Porifera	Sponges: about 10,000 living species	Proterozoic to recent
Coelenterates	Tissues composed of three layers of cells: jellyfish, hydra, coral; about 10,000 living species	Cambrian to recent
Bryozoans	Moss animals: about 3,000 living species	Ordovician to recent
Brachiopods	Two asymmetrical shells: about 260 living species	Cambrian to recent
Molluska	Straight, curled, or two symmetrical shells: snails, clams, squids, ammonites; about 70,000 living species	Cambrian to recent
Annelids	Segmented body with well-developed internal organs: worms and leeches; about 7,000 living species	Cambrian to recent
Arthropods	Largest phylum of living species with over 1 million known: insects, spiders, shrimp, lobsters, crabs, trilobites	Cambrian to recent
Echinoderms	Bottom dwellers with radial symmetry: starfish, sea cucumbers, sand dollars, crinoids; about 5,000 living species	Cambrian to recent
Vertebrates	Spinal column and internal skeleton: fish, amphibians, reptiles, birds, mammals; about 70,000 living species	Ordovician to recent

as algal mats, stromatolites, and microfilaments in chert. The next stage up the evolutionary scale were the protozoans, including the amoebas, the foraminifers, and the radiolarians, which probably built the first limestone formations. The entire body comprised a microscopic single cell containing protoplasm enclosed by a cell membrane. Only the skeletons of the forams and radiolarians have left a sufficient fossil record (Fig. 53). Bacteria and archea

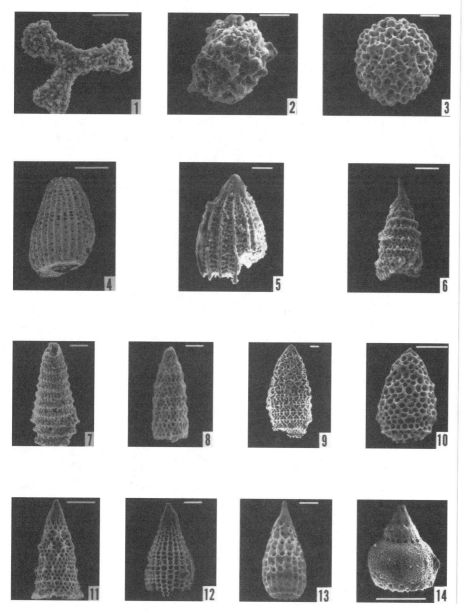

Figure 53 Late Jurassic radiolarians, Chulitna District, Alaska.

(Photo by D. L. Jones, courtesy USGS)

comprise about 4,000 living species; algae and protozoans make up some 80,000 species alive today.

The next step are the sponges, which belong to the phylum Porifera and were the first multicellular animals. They were attached to the ocean floor and grew up to three feet wide. Their bodies were composed of an outer and an inner layer of cells separated by a jellylike protoplasm. The body walls of sponges are perforated by pores, through which water is carried into the central cavity and expelled through one or more larger openings for feeding. Generally, because the sponges were soft-bodied, they did not fossilize well, except the siliceous spicules that made up their skeletons. Some 10,000 species of sponges exist today.

The coelenterates include the jellyfish, sea anemones, hydra, and coral. The soft-bodied animals did not fossilize well, but the calcareous coral made excellent fossils (Fig. 54) and built impressive formations of limestone. More recent corals are responsible for the construction of barrier reefs and atolls, and they even rival humans in changing the face of the Earth. No fewer than 10,000 species of coelenterates inhabit today's ocean.

The bryozoans, or moss animals, are a strange group of animals that lived in extensive colonies attached to the seafloor and filter-fed on microscopic organisms. They have simple calcareous skeletons in the shape of tiny tubes or boxes. Bryozoan colonies show a considerable variety of shapes, including branching and leaflike to mosslike forms, giving the seafloor the appearance of a carpet. They are important marker fossils for correlating rock formations throughout the world.

The brachiopods, or lampshells, are among the most common fossils, with more than 30,000 species cataloged from the fossil record. They resemble clams and scallops but are not related to them. Although they were plentiful during the Paleozoic, few living species are in existence because they endured much mass extinction through the ages. Most modern brachiopods thrive in shallow waters or in intertidal zones. However, many inhabit the ocean bottom between 150 and 1,500 feet, and some thrive at depths reaching 18,000 feet. Brachiopods make excellent index fossils and greatly aid in correlating rock formations.

The mollusks, which include snails, clams, and squids, are a large, diverse phylum that has probably left the most impressive fossil record of all. The three major groups are the clams, snails, and cephalopods. Some forms have a unique mode of transpiration, expelling jets of water. The extinct ammonoids left a large variety of fossil shells. Some had giant spiral shells up to several feet in diameter; others had straight shells over 12 feet long. As many as 70,000 species of mollusks inhabit the world today.

The annelids are segmented worms, whose body is characterized by a repetition of similar parts in a long series. The group includes marine worms,

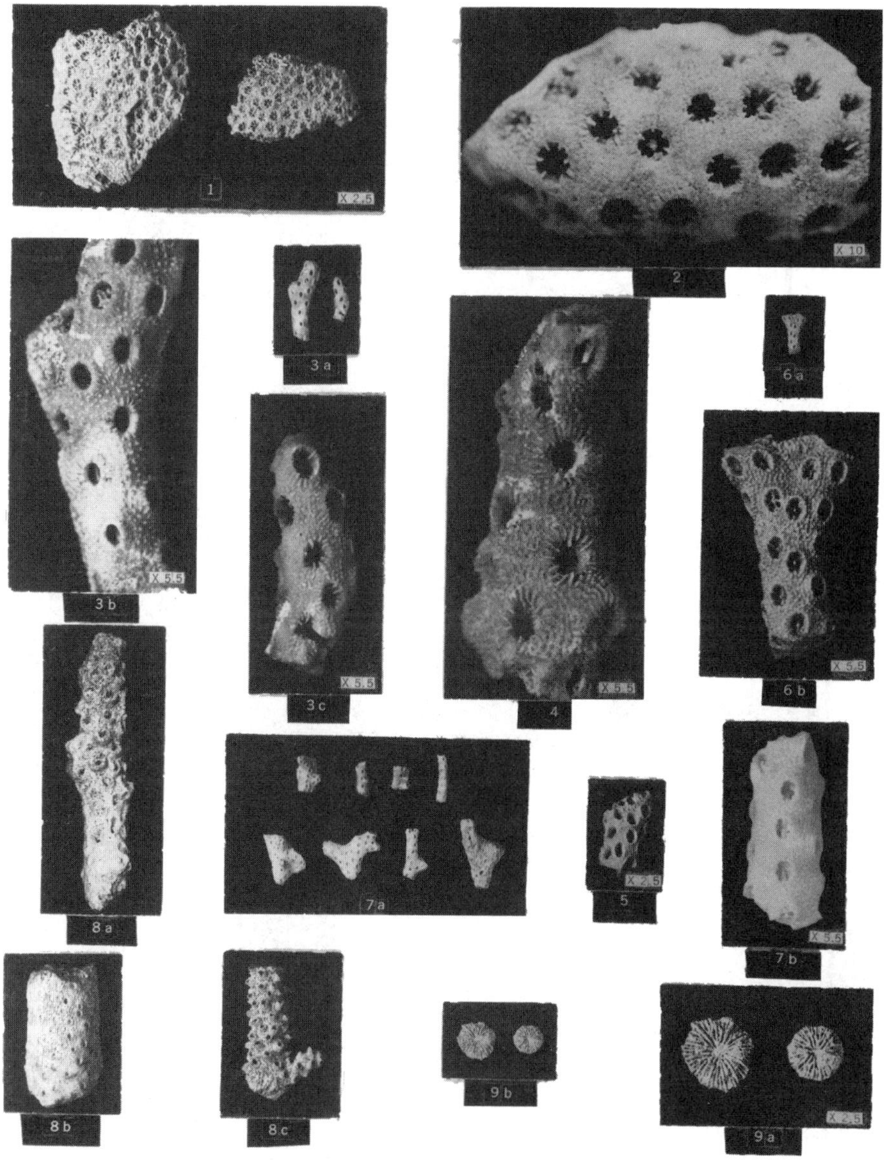

Figure 54 Fossil corals
from Bikini Atoll,
Marshall Islands.

(Photo by J. W. Wells, courtesy USGS)

earthworms, flatworms, and leeches. Marine worms burrowed in the bottom sediments or were attached to the seabed and lived in tubes composed of calcite or aragonite. Because of their soft bodies, worms did not fossilize well; however, they did leave a profusion of tracks and borings, which were often well preserved (Fig. 55). The prolific worms are represented by nearly 60,000 living species.

The arthropods are the largest group of marine and terrestrial invertebrates; they include crustaceans, arachnids, and insects. The marine group includes shrimp, lobsters, barnacles, and crabs; the land group includes insects, spiders, scorpions, daddy longlegs, mites, and ticks. The arthropods have a segmented body and jointed appendages covered with an exoskeleton composed of chitin that must be molted to accommodate growth. The crustaceans comprise about 40,000 living species; insects and related groups hold the record at over 1 million species. Perhaps one of the first and among the best known arthropods were the extinct trilobites, whose fossils make prized possessions.

Figure 55 *Fossil worm borings in the Heiser Sandstone, Pensacola Mountains, Antarctica.*

(Photo by D. L. Schmidt, courtesy USGS)

Figure 56 *Paleontologists unearthing large fossil bones near Littleton, Colorado.*

(Photo by J. R. Stacy, courtesy USGS)

The echinoderms were possibly the strangest animals ever preserved in the fossil record. The group embraces a wealth of extinct forms, surpassing any other phylum. The echinoderms are unique among animals by possessing radial symmetry with body parts radiating from a central point, a water vascular system for feeding and locomotion, and no head. Included are the sea lilies, sea cucumbers, starfish, brittle stars, and sea urchins. Fossils of ancient crinoids and blastoids are eagerly sought after by fossil hunters. Up to 10,000 living species occupy the ocean depths.

The higher animals are the chordates, which include the vertebrates, animals with backbones such as fish, amphibians, reptiles, birds, mammals, and humans. Today, fish comprise about 22,000 species, reptiles about 10,500 species, and mammals about 4,500 species. The vertebrates were the first advanced animals to populate the land. Dinosaur bones, representing some 500 genera, have intrigued scientists since their first discovery in Great Britain in the early 19th century. Today, their popularity is apparent by the large numbers of visitors to natural history museums and of serious patrons of paleontology, who volunteer their time to help paleontologists excavate fossil bones (Fig. 56).

In the plant kingdom, the thallophytes include algae, fungi, and lichens. They have a soft, nonwoody structure that lacks a vascular circulatory system and grow in or near water or in moist places. Only algae possess chlorophyll, which enables them to manufacture nutrients from water, carbon dioxide, and sunlight. The thallophytes generally reproduce by the fusion of male and female single-celled gametes. The earliest fossils are one-celled bacteria and blue-green algae in Precambrian rocks. Fungi and lichens comprise as many as 100,000 living species.

The bryophytes, including mosses and liverworts, were the first plant phylum to become well established on land. They have stems and simple leaves but lack true roots or vascular tissues to conduct water to the higher extremities and therefore are required to live in moist environments. They reproduce by spores that are carried by the wind for wide distribution. The earliest species occupied freshwater lakes in the late Precambrian.

The pteridophytes, or ferns, were the first plant phylum to develop true roots, stems, and leaves. Some present-day tropical ferns grow to tree size as they did in the geologic past. The whisk ferns, which appeared at the end of the Silurian and became extinct near the end of the Devonian, probably gave rise to the first club mosses, horsetails, and true ferns. The true ferns (Fig. 57) are the largest group, both living and extinct, and contributed substantially to Carboniferous coal deposits. Most reproduced with spores, but the extinct seed ferns bore seeds.

The spermatophytes are the higher plants that produce seeds and include the gymnosperms and angiosperms. Gymnosperms are conifers that bear seeds on exposed scales or cones. They ranged from the Carboniferous to the present and covered vast areas in harsh climates. Angiosperms are flowering plants, whose seeds often develop in a fruit. They originated in the Cretaceous and range in size from grasses to huge trees. The higher plants are represented by about 270,000 modern species, although they are presently going extinct at an alarming rate as a result of deforestation in the tropics.

THE SEDIMENTARY ENVIRONMENT

The surface of the Earth both above and below the sea is covered by a relatively thin layer of sediment, and sedimentary rocks are encountered more frequently than any other rock type. They give us not only impressive scenery from ragged mountains to jagged canyons but also much of the wealth of the world. The sedimentary environment provides the conditions necessary for the making of fossils, which supply important clues about the history of the planet. The constant shifting of sediments on the surface and the accumulation of deposits on the ocean floor assure that the face of the Earth will continue to change with time.

Figure 57 Fossil leaves of the tree fern Neuropteris, *Fayette County, Pennsylvania.*

(Photo by E. B. Hardin, courtesy USGS)

Fossils of extinct organisms can be found in most sedimentary rocks, especially limestones and shales. Outcrops of marine sediments are generally the best sites for prospecting for fossils. In most parts of the world, the central portions of the continents were inundated during various times in the geologic past, and thick deposits of marine sediments accumulated in the basins of inland seas. Even the presently high continental interiors were once invaded by inland seas. When the seas departed and the land was uplifted, erosion exposed many of these marine sediments.

Limestones are among the best suited rocks for the preservation of fossils because of the nature of their sedimentation, often involving shells and skeletons of dead marine life that were buried and fused into solid rock. Most limestones originated in marine environments, and some were deposited in lacustrine or lake environments. Limestones are relatively abundant, constituting roughly 10 percent of all exposed sedimentary rocks, and form easily identifiable outcrops (Fig. 58).

Figure 58 *Limestones of the Hawthorn and Ocala formations, Marion County, Florida.*

(Photo by G. H. Espenshade, courtesy USGS)

Most limestones contain whole or partial fossils, depending on whether they were deposited in quiet or agitated waters. Tiny spherical grains called oolites are characteristic of agitated water, whereas lithified layers of limy mud called micrite are characteristic of calm water. In quiet waters, undisturbed by waves and currents, whole organisms with hard body parts are buried in the calcium carbonate sediments and are lithified into limestone.

Shales and mudstones, which commonly contain fossils, are the most abundant sedimentary rocks because they are the main weathering products of feldspars, the most abundant minerals. Because clay particles sink slowly in the sea, they normally settle out in calm, deep waters far from shore. Compaction squeezes out the water between sediment grains, and the clay is lithified into shale. Organisms caught in the clay are compressed into thin carbonized remains or impressions. Fossilization in this manner also can preserve soft body parts much better than other sediments.

Most sedimentary rocks that were deposited in the ocean contain the fossils that record much of Earth history. Because the majority of marine sediments consist of material washed off the continents, most sedimentary rocks form along continental margins or in the basins of inland seas, such as the sea that invaded the interior of North America during the Jurassic and Cretaceous periods. High sediment rates form deposits up to hundreds or even thousands of feet thick. In many places, individual sedimentary beds can be traced for hundreds of miles.

The process of formation of sedimentary rock begins with erosion. Rain plays a significant role in causing erosion, as do wind and glacial ice. Raindrops impart kinetic energy to the soil when they land, kicking up sediment grains and redistributing them farther down slope. Rainwater that does not infiltrate into the ground runs off elevated areas, carrying sediment along with it. Rivulets carry sediment to streams, which transport it to rivers, which in turn dump it into the ocean. Rivers, for example the Mississippi, carry an enormous amount of sediment, which when deposited into the Gulf of Mexico contributes to the continuous outward building of the Gulf Coast region.

Every year, the continents receive about 25,000 cubic miles of rainwater, of which almost half runs back into the sea. An estimated 25 billion tons of sediment, much of which results from soil erosion (Fig. 59), is carried by runoff into the ocean, where it settles out on the continental shelf. The con-

Figure 59 *Severe soil erosion on a farm in Shelby County, Tennessee.*

(Photo by Tim McCabe, courtesy USDA-Soil Conservation Service)

tinental shelf extends up to 100 miles or more and reaches a depth of roughly 600 feet. In most places, the continental shelf is nearly flat with an average slope of only about 10 feet per mile.

By comparison, the continental slope extends to an average depth of two miles or more and has a very steep angle of two to six degrees, comparable to the slopes of many mountain ranges. Sediments that reach the edge of the continental shelf slide down the continental slope under the influence of gravity. Often, huge masses of sediment cascade down the continental slope by gravity slides, which have been known to bury transcontinental telegraph cables under thick deposits of rubble.

Loose sediment grains are also carried by the wind, especially in dry regions, where dust storms are prevalent. The finer particles can remain suspended in the air for long periods. The sediment load in the lower atmosphere is as much as 150,000 tons per cubic mile of air. Particularly strong dust storms such as those in the African Sahara Desert can transport fine-grained sediment over vast distances, even across the Atlantic Ocean to South America.

Fine sediments that land in the ocean slowly build up deposits of abyssal red clay, whose color signifies its terrestrial origin. But most wind-blown sediments remain on the land, where they form thick layers of loess, a fine-grained, sheetlike deposit (Fig. 60). Along with these are dune deposits, composed of desert sands, which when lithified show a distinct dune structure on outcrops, consisting of cross-stratification of sand layers. In addition, the sediment grains of desert deposits are frosted and larger rocks show a desert varnish caused by wind abrasion, similar to sand blasting, and deposits of mineral solutions exuded from within the rock.

Fluvial, or river, deposits are terrestrial sediments that remain, for the most part, on the continent after erosion. When rivers become clogged with

Figure 60 *An exposure of loess standing in vertical cliffs, Warren County, Mississippi.*

(Photo by E. W. Shaw, courtesy USGS)

Figure 61 Close view of cross-bedding in the coarse sands of the Glenns Ferry Formation, Elmore County, Idaho.

(Photo by H. E. Malde, courtesy USGS)

sediments and fill their channels, they spill over onto the adjacent land and carve out a new river course. Thus, rivers meander along downstream, forming thick sediment deposits in broad floodplains that can fill an entire valley. Floodwaters rapidly flowing out of dry mountain regions carry a heavy sediment load, including blocks the size of automobiles. When the stream reaches the desert, its water rapidly percolates into the desert floor, and sometimes huge monoliths dot the landscape, as monuments to the tremendous power of water in motion.

Fluvial deposits are recognized in outcrops by their coarse sediment grains and cross-bedding features (Fig. 61), which were produced when the stream meandered back and forth over old river channels. River currents also can align mineral grains and fossils, giving rocks a linear structure that can be used to determine the direction of current flow. Current ripple marks on exposed surfaces (Figs. 62, 63) can also be used for determining the direction of flow, which is perpendicular to the crest line and toward the angle of least slope, as in present-day streams.

River deposits called alluvium accumulate as a result of a decline in stream gradient, a reduction in stream flow, or a decrease in stream volume, and the heaviest material settles out first. These changes in the river environment occur during entrance into standing water, encounters with obstacles, evaporation, and freezing. River deposition is divided into deposits in bodies of water, alluvial fans, and deposits within the stream valley itself. A medium-size river will take about a million years to move its sandy deposits just 100

Figure 62 Ripple marks
on Dakota Sandstone,
Jefferson County,
Colorado.

(Photo by J. R. Stacy,
courtesy USGS)

miles downstream. Along the way, the grains of sand are polished to a high
gloss, which aids in identifying ancient river deposits. River-deposited sedimentary rocks within streambeds are relatively rare because rivers deliver most

Figure 63 Ripple marks
showing river current
direction.

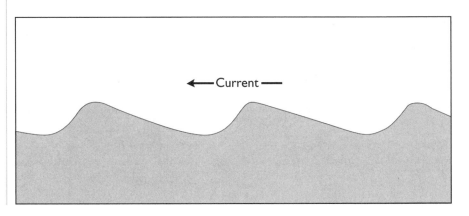

of their sediment load to lakes or the sea. River deltas develop where rivers enter larger rivers or standing water. The velocity of the river slows so abruptly when entering a body of water that its bed load immediately drops out of suspension. Much of the river's load is also reworked by offshore currents, creating marine or lake deposits.

A large portion of the upper midwestern and northeastern parts of the United States, generally above the Missouri and Ohio River valleys, was covered by thick glaciers during the last ice age. Many areas were eroded down to the granite bedrock, erasing the entire geologic history of the region. The power of glacial erosion is well demonstrated by the deep-sided valleys carved out of mountain slopes (Fig. 64) by thick sheets of ice. The glacially derived sediments covered much of the landscape, burying older rocks under thick layers of till.

The glaciers left behind a bizarre collection of structures, including cirques, potholes, kettle holes, glacial lakes, flood-ruptured ground, and many other landforms carved by ice. The glaciers laid down thick deposits of glacial sediment that buried older rocks, forming elongated hillocks aligned in the same direction called drumlins. Long sinuous sand deposits called eskers were formed by glacial debris from outwash streams. Glacial varves, alternating layers of silt and sand in ancient lake bed deposits, were deposited annually in lakes positioned below the outlets of glaciers.

Figure 64 Glacial valley and lake in the Rocky Mountains.

(Courtesy National Park Service)

Lacustrine, or lake environment, deposits are stratified similarly to marine deposits but on a smaller scale, depending on the size of the lake. Glacial lakes, such as the Great Lakes, which were huge pits carved out by the glaciers and later filled with meltwater, receive large amounts of sediments derived from the continent. The buildup of sediments continually causes the lakes to become shallower, until at some time in the future they will dry out completely and become flat, featureless planes.

Many large lakes or inland seas in the interiors of the continents are salt-water lakes similar to the Great Salt Lake in Utah. Some saltwater lakes might have originated as pinched-off sections of retreating seas that once invaded the continent. Because they generally have no outlet, the lakes become increasingly salty as rivers continue to carry salts into them. The Great Salt Lake is eight times saltier than the ocean, giving swimmers greater buoyancy. The Dead Sea on the border between Israel and Jordan became so salty that it completely overturned, with the bottom waters rising to the surface, killing off most of its plant and animal life.

FOSSIL PRESERVATION

The branch of geology devoted to the study of ancient life based on fossils is called paleontology. Fossils are the remains or traces of organisms preserved from the geologic past. Not all organisms become fossils, however, and plants and animals must be buried under certain specific conditions to become fossilized. Given enough time, the remains of an organism are modified, often becoming petrified: literally turned to stone.

Paleontologists have assembled a fairly good picture of ancient life on this planet, because living things tend to record their own history. The story of ancient life is written in the fossil record of extinct organisms, which leave their epitaphs inscribed on stone. The term *fossil* comes from the Latin *fossilis,* meaning "to dig," as paleontologists and amateur fossil hunters often have to do to find them. The rarest as well as the most interesting fossils are the actual remains of organisms, which include bones, shells, and other well-preserved body parts.

Fossils are formed in a variety of ways under many different environmental conditions. In rare circumstances, the complete animal is preserved. More frequently, the durable parts of the organism are preserved in a relatively unaltered condition. But mostly, fossils are extensively altered so that little of the original material remains, although the actual shapes and textures are recognizable. The greatest number of fossils comprise not the remains of organisms themselves but indirect evidence of their existence by a preponderance of tracks, trails, burrows, and imprints.

Generally, in order to become preserved in the fossil record, organisms must possess hard body parts such as shells or bones. Soft fleshy structures are quickly destroyed by predators or decayed by bacteria. Even hard parts left on the surface for any length of time will be destroyed. Therefore, organisms must be buried rapidly to escape destruction by the elements and to be protected against agents of weathering and erosion. Marine organisms thus are better candidates for burial than those living on the land because the ocean is typically the site of sedimentation, whereas the land is largely the site of erosion.

The beds of ancient lakes were also excellent sites for rapid burial of skeletal remains of freshwater organisms and skeletons of other animals, including those of early humans. Ancient swamps had prolific growth of vegetation, which fossilized in abundance. Many animals became trapped in bogs overgrown by vegetation. The normally reducing environment of the swamps kept bacterial decay to a minimum, thereby greatly aiding the preservation of plants and animals. The rapidly accumulating sediments in floodplains, deltas, and stream channels buried freshwater organisms, along with other plants and animals that happened to fall into the water.

Only a small fraction of all the organisms that have ever lived are preserved as fossils. Normally, the remains of a plant or animal are completely destroyed through predation and decay. Although becoming a fossil does not seem very difficult for some organisms, for others it is almost impossible. For the most part, the remains of organisms are recycled in the Earth; that is fortunate because soil and water would soon become depleted of essential nutrients. Also, most of the fossils exposed on the Earth's surface are destroyed by weathering processes. This makes for an incomplete fossil record with poor or no representation of certain species.

The best fossils are those composed of unaltered remains. The more durable parts of some organisms as old as 70 million years or more have been preserved in their original composition and appearance. Usually the remains of plants and animals must be buried quickly in order to form a fossil. Otherwise, they will be eaten by scavengers, decayed by bacteria, or broken up and scattered by wind or water. Generally, the inorganic hard parts, composed mostly of calcium carbonate, form the vast majority of unaltered fossils. Calcite and aragonite also contribute to a substantial number of fossils of certain organisms. To a lesser extent, calcium phosphate, which constitutes the bones and teeth of most vertebrates, remains unaltered. In addition, entire bodies might be sealed in a protective medium such as tree sap, tar, or glacial ice.

The next best fossils comprise altered remains. The original substance, whether wood or bone, is replaced by minerals from circulating solutions, or pore spaces are filled with minerals and the specimen is literally turned to stone. When the skeleton of an animal is buried in the sediments, the original cellular materials are gradually lost as mineral-laden water soaks into the cell spaces. As the water evaporates, minerals are left behind as a layer of solid material lining each cell. Eventually, after repeated soaking and drying, the original cell is completely replaced by mineral matter in the exact same pattern as the bone.

The usual petrifying agents carried by the groundwater are calcite and silica. These minerals originate when atmospheric carbon dioxide reacts with rainwater to produce a weak solution of carbonic acid, which leaches minerals from the soil and filters down to the water table. If iron sulfide is used, the original material is replaced with pyrite crystals, providing a beautiful and unique specimen. The most common petrified fossils are large dinosaur bones, which can be quite heavy, and tree trunks such as those found in the Petrified Forest of Arizona (Fig. 65).

If groundwater dissolves the remains buried in the sediment, a mold is left behind. The mold faithfully reflects the shape and surface markings of the organism; however, it does not reveal any information about its internal structure. When the mold is subsequently filled with mineral matter, a cast is formed (Fig. 66). The brain size of fossil animals can be determined by mea-

Figure 65 *Petrified logs at the Petrified Forest National Park, Apache County, Arizona.*

(Photo by Richard Frear, courtesy National Park Service)

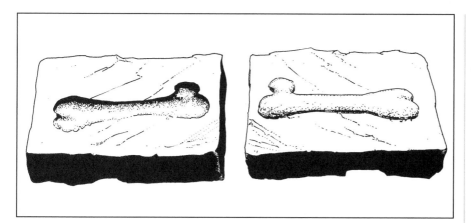

Figure 66 *Comparative views of a mold (left) and a cast (right). A cast is formed when a hollow mold is filled with mineral matter.*

suring the volume of casts, called endocasts, made from the animal's skull, which are exact replicas of the cranial cavity.

The remains of plants and some small animals are altered by a process of carbonization in coal deposits. The coal beds formed when trees and other plants accumulated in swamps and began to decompose. After most volatile substances were driven off, cellular materials such as cellulose were gradually changed by the concentration of carbon, resulting in coal. Between easily split layers of coal or shales separating coal seams are readily recognizable leaves and stems of plants.

Leaves and delicate animal forms are also preserved by carbonization, when fine sediment such as clay encases the remains of an organism. As time passes, pressure from overlying rocks squeezes out the liquid and gaseous components and leaves behind a thin film of carbon. Black shales deposited as organic-rich mud in oxygen-poor environments such as swamps often contain abundant carbonized remains. If the carbon film is lost from a fossil preserved in fine-grained sediment, a replica of the surface of the plant or animal, called an impression, might still show considerable detail.

Other delicate organisms, such as insects, are difficult to preserve in this manner. Consequently, they are quite rare in the fossil record. Not only do they need to be protected from decay, but they must not be subjected to any pressure that could crush them. Insects and even small tree frogs might be preserved in amber, the hardened resin or sap of ancient trees. The animal first becomes trapped in the sticky resin, which seals it off from the atmosphere and protects its remains from damaging air and water. When the resin hardens, it forms a protective pressure-resistant case around the insect. Air is also sometimes trapped in amber. Gas analysis of air bubbles inside Cretaceous age amber about 80 million years old indicates that the atmosphere might have had a substantially higher oxygen content than it does today, perhaps accounting in part for the gigantism of the dinosaurs.

Some types of fossils are not the remains of the animal itself, but what it left behind. Some dinosaur species swallowed gizzard stones, similarly to modern birds, in order to grind the vegetation in their stomachs into pulp. The rounded, polished stones called gastroliths were left in a heap where the dinosaur died, and often deposits of these stones can be found on exposed Mesozoic sediments of the American West.

Coprolites are masses of fecal matter preserved as fossils and are usually modular, tubular, or pellet-shaped. Coprolites often are used to determine the feeding habits of extinct animals. For example, coprolites of herbivorous dinosaurs are black and block-shaped and usually filled with plant material. Those of carnivorous dinosaurs are spindle-shaped and contain broken bits of bone consumed while dining on other animals. Tunnels bored through the fossilized feces also record the earliest dinosaur dung-eating beetles. Dinosaur coprolites can be quite massive, even larger than a loaf of bread.

Pseudofossils (false fossils) resemble natural fossils in appearance but were not made by living things. Some are crystalline aggregates of other accumulations of mineral matter of inorganic origin called concretions. They are lumps or nodules found in shales, sandstones, and limestones, the same rocks where fossils are found. Some concretions actually do contain fossil shells, fish, or insects. Many concretions look like petrified dinosaur eggs because of their similar rounded shapes. Others have cracks filled with minerals and are often mistaken for fossil turtle shells.

TRACKS, TRAILS, AND FOOTPRINTS

The bones of extinct animals are much rarer than their footprints, and many animals are known only by their tracks. The formation of clear foot impressions requires a moist, fine-grained, and cohesive sediment bed for the animal to walk on (Fig. 67). If the animal walks slowly, it will leave a detailed impression of its feet. Even clear outlines of claws or nails, the shape of the footpad, and the pattern of scales can be discerned.

Unfortunately, few such high-quality fossil footprints are found, and most are partially destroyed during the sedimentary process that buries and preserves them. The most favorable conditions for the preservation of footprints exist after high tide waters have receded, and the tracks are allowed to dry and harden and eventually fill with a different type of sediment. The weight of the animal is also important; large animals, such as dinosaurs, leave deeper tracks that are less readily destroyed and most likely to be preserved.

Measurements of trackways can determine the stride of the animal and sometimes even its pace, whether walking or running. Dinosaur tracks provide clear fossil footprints because the great weight of many species produced deep

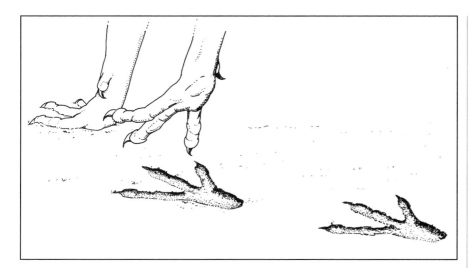

Figure 67 *Formation of foot impressions requires a moist, fine-grained sediment bed for the animal to walk on.*

indentations in the ground. Their footprints exist in relative abundance in terrestrial sediments of Mesozoic age throughout most of the world. The study of dinosaur tracks suggests that some species were highly gregarious and gathered in herds. Large carnivores such as Tyrannosaurus rex were swift, agile predators that could sprint at speeds up to 20 miles per hour or more according to their tracks.

Animal tracks tell of the earliest invasion on dry land some 370 million years ago. Primitive Devonian fish, similar to today's lungfish, crawled on their bellies from one pool to another, using lobed fins to push themselves along. The lobed-finned fish that first ventured on land gave rise to the four-legged amphibians, whose tracks are found in formations of late Devonian age. The amphibian tracks are generally broad and have a short stride, indicating that the animal could barely hold its squat body off the ground. It walked with a clumsy gait, and running was out of the question. Amphibian footprints became abundant in the Carboniferous period but less so in the Permian, as a result of the ascendancy of the reptiles and the amphibians' preference for life in the water.

The increase in the number of reptilian footprints in the Carboniferous and Permian plainly shows the rise of the reptiles at the expense of the amphibians. Possibly one of the major factors leading to the superiority of the reptiles was their more efficient mode of locomotion. The reptiles were also much more suited for living full time on dry land, whereas the amphibians had to return to the water to moisten their skin and to reproduce. Although most reptiles walked or ran on all fours, by the late Permian, some smaller species reared up on their hind legs when they wanted to move swiftly. Their body pivoted at the hips and their long tail counterbalanced their nearly erect trunk. This stance probably freed their forelimbs for attacking prey. At the beginning of the Mesozoic era, about 250 million years ago, the dinosaurs descended from the thecodonts, which were two-legged reptiles, and many early dinosaurs developed a successful permanent bipedal stance. This increased their speed and agility and freed the forelimbs for foraging and performing other useful tasks that clumsy legs were unable to do. The back legs and hip therefore had to support the entire weight of the animals, probably preventing them from growing larger than they did. Apparently as a result of increased weight, some bipedal dinosaurs later reverted to a four-legged stance, which eventually gave rise to the gigantic apathosaurus.

Dinosaur tracks are the most spectacular of all fossil footprints (Fig. 68), and they are found in relative abundance in terrestrial sediments of Mesozoic age in most parts of the world. Indeed, the climate during this time was mild enough to allow the migration of dinosaurs to just about every corner of the globe. Through time, the tracks of many dinosaurs show a lengthening of the stride, a narrowing of the trackway, and forward pointing of the toes, all indications of increasingly efficient and fast locomotion. Some tracks were nearly mammallike in their structure, possibly indicating an intermediate species between reptile and mammal. By the end of the Mesozoic, dinosaur footprints disappeared entirely from the face of the Earth to be replaced by a preponderance of mammal tracks.

When the dinosaurs became extinct, the mammals were poised to take over the world. At first, the majority of mammals were small, nocturnal creatures with a limited range. Although they probably left numerous tracks, their light weight produced shallow footprints that did not fossilize well. As the mammals became larger, however, they left a better fossil record of their

Figure 68 *A fossil dinosaur footprint in Dakota sandstone, Jefferson County, Colorado. The formation of a fossil footprint requires deep, distinct impressions, which are later filled with sediment and lithified into solid rock.*

(Photo by J. R. Stacy, courtesy USGS)

comings and goings. The stride of the animals lengthened, the trackway narrowed, and the digits were directed forward, indicating an increase in efficiency and speed.

In 1976, at Laetoli, Tanzania, well-preserved footprints of a large mammal were found embedded in a volcanic ash bed that was dated at 3.8 million years old. The footprints had rounded heels and arches, pronounced balls, of the feet, and forward-pointing toes, all indications of an efficient form of locomotion. In addition, the tracks clearly showed that the animal preferred a bipedal walk. The most astonishing fact was that these fossil footprints were from one of our ancient ancestors, who walked upright much earlier than ever thought possible.

The next two chapters will take a closer look at some of the more fascinating marine and terrestrial fossils.

5

MARINE FOSSILS
CREATURES PRESERVED FROM THE SEA

The discovery, reliable dating, and piecing together of marine fossils have provided us with a fascinating window into the earliest forms of animal life on Earth. However, some disagreement has occurred over the placement of certain obscure organisms into existing taxonomic divisions. Paleontologists are generally divided into two groups of people—lumpers and splitters. Lumpers like to place all types of similar organisms into a single taxonomic category, whereas splitters prefer to distinguish many separate groupings.

Some classification schemes place the protozoans in the kingdom of Protistae, which includes all single-celled plants and animals that possess a nucleus. In the obscured past, few distinctions existed among the earlier plants and animals, and they shared similar characteristics. Other classification schemes prefer to place the protozoans squarely in the animal kingdom, and indeed, *protozoan* literally means "beginning animal."

PROTISTIDS

The protistids were the first group of organisms to evolve a nucleus. Some varieties of protistids formed large colonies, whereas most lived independently. The entire body is composed of a single cell containing living protoplasm enclosed within a membrane. Present-day single-celled organisms have not changed significantly from the fossil forms. Most, however, were soft-bodied and did not fossilize well. They obtained energy by ingestion of food particles or by photosynthesis. Reproduction is by the use of unicellular gametes that do not form embryos.

The ability to travel under their own power is what essentially separates animals from plants. Some unicellular animals moved about with a thrashing tail, called a flagellum. It resembled a filamentous bacteria that combined with the single-celled animal in a symbiotic relationship for mutual benefit. Other cells had tiny hairlike appendages, called cilia, which helped them travel by rhythmically beating the water. Amoeba had a unique form of transportation: extending fingerlike protrusions outward from the main body and flowing into them.

The major fossil groups of protistids are algae, diatoms, dinoflagellates, radiolarians, foraminifers, and fusulinids. Among the earliest protistids were microorganisms that built stromatolite structures (Fig. 69). Ancestors of blue-green algae built these concentrically layered mounds, resembling cabbage heads, cementing sediment grains together by using a gluelike substance secreted from their bodies. As with modern stromatolites, the ancient stromatolite colonies grew in the intertidal zone, and their height, which was as much as 30 feet, was indicative of the height of the tides during their lifetime. This is because stromatolites grow between the low-tide water mark and the high-tide mark.

Some varieties of protistids formed large colonies, but most lived independently. The organisms ranged from the Precambrian to the present, although many did not become well established until the Cambrian or later. Because of their lack of a hard shell, the amoeba and paramecium did not fossilize well. The radiolarians, which are well represented in the fossil record, usually have siliceous shells of remarkably intricate designs, including needle-like, rounded, or an open network structures of delicate beauty. Different species of fossil radiolarians are used to determine the ages of oceanic crustal blocks called terranes that accrete to the edges of continents and also define specific regions of the ocean where the terranes originated.

The foraminifers (Fig. 70) were microscopic protozoans, whose skeletons composed of calcium carbonate preserved much of the record of the behavior of the ocean and climate. Most lived on the bottom of shallow seas; a few floating forms existed as well. Their remains are found in both shallow and deep water deposits. The prolific forams are particularly useful to petroleum geologists, who rely on them for dating the stratigraphy of oil well cuttings. The fusulinids were large, complex protozoans that resembled grains of wheat, ranging from microscopic size to giants up to three inches in length.

Figure 69 *Stromatolite structure near the junction of Canyon Creek and Salt River, Gila County, Arizona.*

(Photo by A. F. Shride, courtesy USGS)

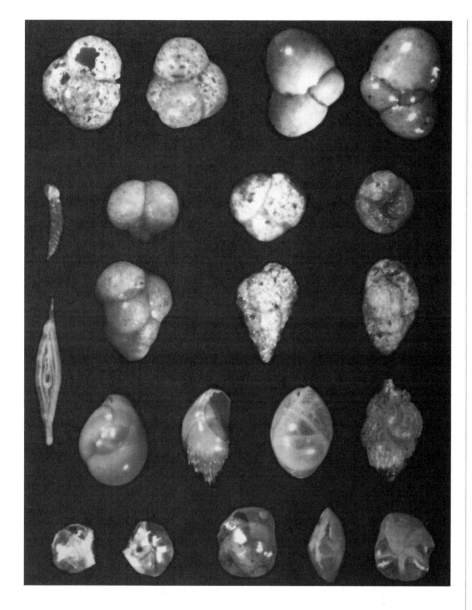

Figure 70 *Foraminifera from the North Pacific Ocean.*

(Photo by R. B. Smith, courtesy USGS)

Although individually, the protistids did not leave fossils that are of any interest to most amateur collectors, some varieties did perform a major function by building massive formations of limestone, which entombed other fossils. When the tiny organisms died, their shells fell to the ocean floor in a constant rain. The shifting of these sediments by storms and undersea currents buried dead marine organisms that were not eaten by scavengers. A calcite ooze then formed and eventually hardened into limestone, preserving trapped species through the ages.

SPONGES

As time progressed, individual cells joined to form multicellular animals called metazoans, which evolved in the latter part of the Proterozoic, around 600 million years ago. The metazoans gave rise to ever more complex organisms, which were ancestral to all marine life today. The first metazoans were a loose organization of cells united for a common purpose such as locomotion, feeding, and protection. If cells became separated from the main body, they could exist on their own until they either regrouped or grew separately into mature species.

The most primitive metazoans were probably composed of a large aggregate of cells, each with its own flagellum. The cells congregated into a small, hollow sphere, and their flagellum rhythmically beat the water to propel the tiny creature around. Some metazoans were literally turned inside out and attached themselves to the ocean floor. They had openings to the outside, and the flagella, now on the inside, produced a flow of water that carried food particles in and wastes out. These were probably the forerunners of the sponges.

The sponges presented various shapes and sizes and often grew in thickets on the ocean floor. Some early species grew to enormous size: 10 feet or more across (Fig. 71). The body consisted of three weak tissue layers, whose cells were still capable of independent survival. If a sponge were cut up into tiny pieces, each individual part could grow into a new sponge. The body walls are perforated by pores, through which water is carried into the central cavity and expelled through one or more larger openings.

Certain sponge types have an internal skeleton of rigid, interlocking spicules composed of calcite or silica. One group had tiny glassy spikes for

Figure 71 The sponges were the first giants of the sea, growing 10 feet or more across.

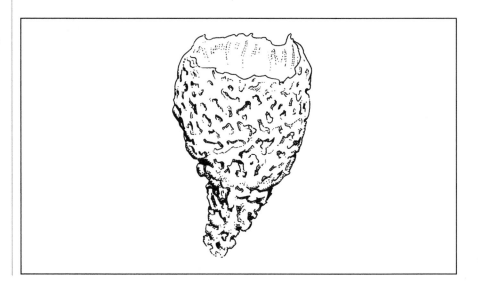

spicules, which gave the exterior a rough texture unlike that of their smoother relatives used in the bathtub. The so-called glass sponges consisted of glasslike fibers of silica intricately arranged to form a beautiful network. These hard skeletal structures are generally the only parts of sponges preserved as fossils. The great success of the sponges along with organisms such as diatoms that extract silica directly from seawater to construct their skeletons explains why today's ocean is largely depleted of this mineral. Sponges ranged from the Precambrian to the present, but microfossils of sponge spicules did not become abundant until the Cambrian.

COELENTERATES

An evolutionary step above the sponges were the jellyfish, which belong to the phylum Coelenterata, from Greek meaning "gut." Most coelenterates are radially symmetrical, with body parts radiating outward from a central point. The coelenterate body generally has a saclike shape and a mouth surrounded by tentacles. Primitive, radially symmetrical animals have just two types of cells, the ectoderm and endoderm, whereas the bilaterally symmetrical animals also have a mesoderm (intermediate layer) and a distinct gut. During early cell division in bilateral animals, called cleavage, the fertilized egg forms two, then four cells, each of which gives rise to many small cells.

Jellyfish have two layers of cells separated by a gelatinous substance that gives the saucerlike body a means of support. Unlike those of the sponges, the cells of the jellyfish were incapable of independent survival and were linked by a primitive nervous system that enabled them to contract in unison. These cells thus became the first simple muscles used for locomotion. Jellyfish ranged from the top of the Precambrian to the present, but because they lacked hard body parts, they are rare as fossils and usually are only preserved as carbonized films or impressions.

More advanced than the jellyfish are the corals, which exist in a large variety of forms. Corals are soft-bodied animals that live in individual skeletal cups or tubes called thecae composed of calcium carbonate. Many are well represented in the fossil record, leaving fossils that closely resembled their modern counterparts (Fig. 72). Corals began constructing reefs in the early Paleozoic, forming barrier islands and island chains. The corals also built atolls on top of extinct marine volcanoes, and as the volcanoes subsided beneath the sea, the corals' rate of growth matched the rate of subsidence, keeping them at a constant depth. Reef-building corals created the foundations for spectacular underwater edifices that cover about three-quarters of a million square miles of the Earth's surface and house about a quarter of all marine species. The corals diverged into two basic lineages before their ability to build calcified

Figure 72 *Coral at Bikini Atoll, Marshall Islands.*

(Photo by K. O. Emery, courtesy USGS)

skeletons developed, suggesting they might have evolved a reef-building capability twice in geologic history. The coral polyp is a soft-bodied creature that is essentially a contractible sac, crowned by a ring of tentacles (Fig. 73). The tentacles surround a mouthlike opening and are tipped with poisonous stingers. The polyps live in individual thecae and extend their tentacles to feed at night and withdraw into their thecae during the day or at low tide to protect from drying out in the sun.

Figure 73 *Coral polyps seek protection in carbonate cups from predators and during low tide.*

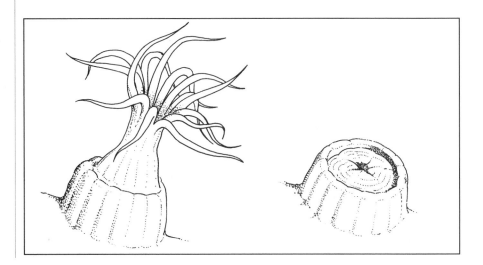

The corals coexist in symbiosis with zooxanthellae algae, which live within the polyp's body. The algae consume the coral's waste products and produce organic materials that are absorbed by the polyp. Some coral species receive 60 percent of their food from algae. Because the algae require sunlight for photosynthesis, corals are restricted to warm, shallow seas generally less than 100 feet deep and normally thrive at temperatures between 25 and 29 degrees Celsius.

Living on the coral reefs is a riot of plant and animal life, more than in any other marine environment. The key to this prodigious growth are the unique biologic characteristics of corals, which play a critical role in the structure, ecologic properties, and nutrient cycles of the reef community. Coral reef environments have the highest rates of photosynthesis, nitrogen fixation, and limestone deposition of all habitats. The most remarkable feature of coral colonies is their ability to build massive calcareous skeletons, weighing several hundred tons.

Small, fragile corals and large communities of green and red calcareous algae live on the coral framework. Hundreds of species of encrusting organisms such as barnacles also thrive on the coral edifice. Large numbers of invertebrates and fish hide in the nooks and crannies of the reef, often waiting until nighttime before emerging to feed. Other organisms attach to practically all available space on the underside of the coral platform or on dead coral skeletons. Filter feeders such as sponges and sea fans occupy the deeper regions.

The archaeocyathans (Fig. 74) resembled both corals and sponges but have no close relationship to any living group and therefore belong to their own unique phylum. They formed the earliest reefs and became extinct in the Cambrian. The tabulate corals, which became extinct at the end of the

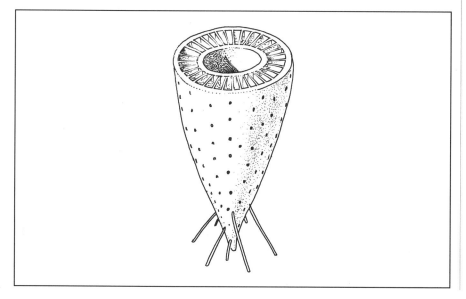

Figure 74
Archaeocyathans built the earliest limestone reefs.

Paleozoic, consisted of closely packed polygonal or rounded tubes, some of which had pores covering their walls. The rugose corals also called horn corals, named so because of their typical hornlike shapes, were the major reef builders of the late Paleozoic and became extinct in the early Triassic. The hexacorals, which ranged from Triassic to recent were the major reef builders of the Mesozoic and Cenozoic eras.

Today, many plant and animal communities thrive on the coral reef because of the coral's ability to build huge wave-resistant structures. Australia's Great Barrier Reef is one of the most impressive natural wonders of the world and the largest feature built by living organisms. In the geologic past, these massive structures were lithified into limestone, creating some of the greatest fossil deposits known on Earth.

BRYOZOANS

The bryozoans are similar in appearance to corals, but are more closely related to brachiopods. They comprised microscopic individuals living in small colonies up to several inches across, giving the ocean floor a mosslike appearance. Like corals, bryozoans are encased in a calcareous vaselike structure, often in the form of small tubes or boxes. They retreat into this structure for protection from predators or from the sun during low tide. Living species occupy seas at various depths, with certain rare members adapted to life in fresh water. They have existed since the Cretaceous about 140 million years ago.

A new colony forms from a single free-moving larval bryozoa that fixes onto a solid object and grows into numerous individuals by a process of budding, which is the production of outgrowths. The polyp has a circle of ciliated tentacles, forming a sort of net around the mouth and used for filtering microscopic food floating by. The tentacles rhythmically beat back and forth, producing water currents that aid in capturing food, which is digested in a U-shaped gut. Wastes are expelled outside the tentacles just below the mouth.

Bryozoan fossils are common in sedimentary rocks, particularly when covering the bedding surfaces of rocks. Fossil bryozoans are abundant in Mississippian formations, especially those of the American Midwest and Rocky Mountains. They resemble modern descendants, with some larger groups possibly contributing to Paleozoic reef building, producing extensive limestone formations. The fossils are most abundant in limestone and less plentiful in shales and sandstones. Often, a delicate outline of bryozoans can be seen encrusting fossil shells of aquatic animals, stones, and or other hard bodies.

The identification of various species is determined by the complex structure of the skeletons, which aids in delineating specific geologic periods. Bryozoans have been very abundant, ranging from the Ordovician to the present, and their fossils are highly useful for making rock correlations. Because of their small size, bryozoans make ideal microfossils for dating oil well cuttings.

BRACHIOPODS

The most common fossils are the brachiopods, also called lampshells because of their resemblance to ancient oil lamps with protruding wicks. The appearance of large numbers of brachiopod fossils in rock formations indicates seas of moderate to shallow depth. Rocks from the Cambrian and Devonian periods contain brachiopod fossils and wave marks, indicating that some ancient forms inhabited the shore areas. Modern forms, which number about 260 species, inhabit warm ocean bottoms from a few feet to over 500 feet deep, and some rare types thrive at depths approaching 20,000 feet.

Primitive forms called inarticulates had two saucerlike shells, or valves, fitted face to face and opened and closed by simple muscles. The shells were not hinged and are often found as separate halves. More advanced species called articulates are similar to those living today. They had ribbed shells of unequal size and shape, which were symmetrical down the center line with interlocking teeth that opened and closed along a hinge line.

The valves are lined on the inside with a membrane called a mantle, which encloses a large central cavity that holds the lophophore, which functions in food gathering. Projecting from a hole in the valve is a muscular stalk called a pedicle by which the animal is attached to the seabed. The structure of the valves aids in the identification of various brachiopod species. The shells take a wide variety of forms, including ovoid, globular, hemispherical, flattened, convex-concave, and irregular. The surface is smooth or ornamented with ribs, grooves, or spines. Growth lines and other structures show changes in form and habit that offer clues to brachiopod history.

The brachiopods frequently were attached to the ocean floor by a footlike appendage and fed by filtering food particles through open shells. Brachiopod shells are often confused with shells of clams (Fig. 75), which are also bivalved but belong to the more advanced mollusks. Clam shells are typically left and right in relation to the body and are mirror images of each other; each valve is asymmetrical down the center line.

Brachiopods ranged from the Cambrian to the present but were most abundant in the Paleozoic and to a lesser extent in the Mesozoic. More than 30,000 species are cataloged in the fossil record. Many brachiopods are excellent index fossils for correlating rock formations throughout the world. They are important as guide fossils and are used to date many Paleozoic rocks.

Figure 75 *A comparison of the shells of brachs (left) and clams (right).*

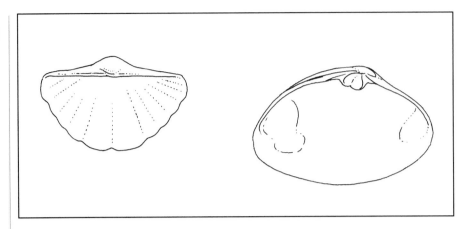

Figure 75 *A comparison of the shells of brachs (left) and clams (right).*

MOLLUSKS

The mollusks are probably the most common and diverse group of animals and left the most impressive fossil record of any phylum (Fig. 76). They make up the second largest of the 21 basic animal groups. The phylum is so diverse paleontologists have difficulty finding common features among its members. The mollusks were well represented in the middle Paleozoic, and the first appearance of freshwater clams suggests that aquatic invertebrates had conquered the land by this time. The mollusks became the most important shelled invertebrates of the Mesozoic seas, and about 70,000 distinct species are living today.

The three major types of mollusks are the snails, clams, and cephalopods. Snails and slugs, which make up the largest group, ranged throughout the Phanerozoic. The mollusk shell is an ever-growing one-piece coiled structure for most species and a two-part shell for clams and oysters. Mollusks have a large muscular foot that is used for creeping or burrowing or is modified into tentacles for seizing prey.

The clams are generally burrowers, although many are attached to the ocean floor. The shell consists of two valves that hang down on either side of the body and, except in scallops and oysters, are mirror images of each other. The cephalopods, which include squid, cuttlefish, octopus, and nautilus, travel by jet propulsion. The animals suck water into a cylindrical cavity through openings on each side of the head and expel it under pressure through a funnellike appendage.

The extinct nautiloids grew to lengths of 30 feet and more and with their straight, streamlined shells were among the swiftest and most spectacular creatures of the deep. The belemnoids were abundant during the Jurassic and Cretaceous and became extinct in the Paleogene. The chambered part of the belemnoid shell was smaller than that of the nautiloids, and the outer walls

Figure 76 *Molds and shells of mollusks on highly fossiliferous sandstone of the Glenns Ferry Formation, Deadman Creek, Elmore County, Idaho.*

(Photo by H. E. Malde, courtesy USGS)

were thickened into a fat cigar shape. The extinct ammonoids (Figs. 77a and b) are the most significant cephalopods and are characterized by a large variety of coiled shell forms which made them ideal for dating Paleozoic and Mesozoic rocks. Some ammonoids grew to tremendous size with shells up to seven feet across, and more awkward types with straight shells attained lengths of 12 feet or more.

ANNELIDS

The annelids are segmented worms, whose body is characterized by a repetition of similar parts in a long series. They include marine worms, earthworms, flatworms, and leeches. Marine worms burrowed in the bottom sediments or were attached to the seabed, living in tubes composed of calcite or aragonite.

The tubes were almost straight or irregularly winding and attached to a solid object such as a rock, a shell, or coral. Early forms of marine flatworms grew very large, nearly three feet long but less than one-tenth-inch thick. The animals were extremely flat in order to absorb directly into their bodies nutrients and oxygen from seawater, present in only small concentrations compared to what they are today.

The primitive segmented worms developed muscles and other rudimentary organs, including sense organs and a central nervous system to process information. The coelomic, or hollow-bodied worms adapted to burrowing in the bottom sediments and might have given rise to more advanced animals. The annelids ranged from the upper Precambrian to the present. Their fossils, which are rare, consist mostly of tubes, tiny teeth and jaws, and a preponderance of fossilized tracks, trails, and burrows.

Soft-bodied marine animals living in the Silurian period some 430 million years ago included a host of worms and bizarre arthropods. More enigmatic yet was a small stubby-legged worm called a lobopod. A large variety of wormlike creatures lived in the Cambrian and apparently evolved into higher forms of animal life. Late in the Cambrian when burrowing shelled marine invertebrates evolved, they often disrupted the corpses of soft-bodied animals buried in the sediment before fossilization could occur, making them extremely rare in the fossil record.

ARTHROPODS

The arthropods are the largest group of animals, living or extinct, and they make the most fascinating fossils. Roughly a million species are alive today,

Figure 77b *A collection of ammonite fossils.*

(Photo by R. B. Smith, courtesy of USGS)

constituting about 80 percent of all known animals. The arthropods conquered land, sea, and air and are found in every environment on Earth. The body of an arthropod is segmented, suggesting a relationship to the annelid worms. Paired, jointed limbs are generally present on most segments and are

Figure 78 Crustaceans such as this crab are primarily aquatic species.

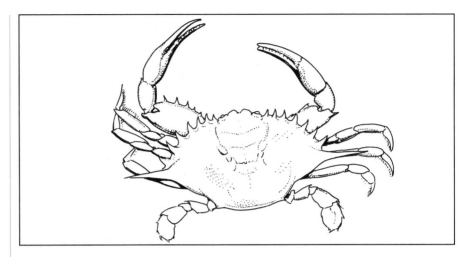

modified for sensing, feeding, walking, and reproduction. One giant arthropod found in the middle Cambrian Burgess Shale Formation of western Canada was as much as three feet long. The crustaceans are primarily aquatic and include shrimp, lobsters, barnacles, and crabs (Fig. 78). Of particular importance to geologists are the ostracods, or mussel shrimp, which are used for correlating rocks from the Ordovician onward.

The arachnids mainly comprise air-breathing species and include spiders, scorpions, daddy longlegs, ticks, and mites. One giant Paleozoic sea scorpion had massive claws and grew over six feet long. The extinct eurypterids (Fig. 79), which ranged from the Ordovician to the Permian, were also giants

Figure 79 The extinct eurypterid, which could grow up to six feet long.

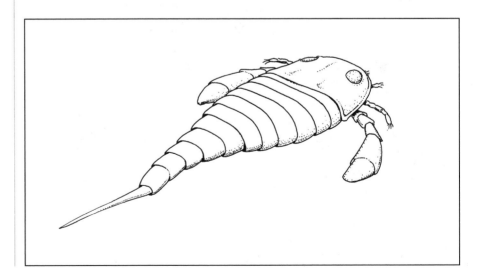

that grew up to six feet long. Its descendants were among the first creatures to go ashore to prey on crustaceans.

The insects are by far the largest living group of arthropods. They have three pairs of legs and typically two pairs of wings on the thorax, or mid-section. To fly, insects had to be lightweight; therefore, their delicate bodies did not fossilize well, except when they were trapped in tree sap, which altered into amber. In most cases, the insect body is covered by an exoskeleton made of chitin, which is similar to cellulose and covers the outside of the body and the appendages. In some groups, the exoskeleton is composed of calcite or calcium phosphate, which greatly improved the animal's chances of fossilization.

The Cambrian is best known for the trilobites (Fig. 80), which appeared at the base of the period and became the dominant species of the early Paleozoic. Because trilobites were widespread and lived throughout the Paleozoic, their fossils are important markers for dating rocks of that era. Trilobites are thought to be primitive ancestors of the horseshoe crab, the only remaining direct descendant alive today. The giant paradoxides extended near-ly two feet in length, compared to most trilobites, which were one-half inch to four inches long.

The trilobite is divided into three lobes (hence its name), which include the head, body, and abdomen. Because trilobites shed their exoskeleton as they grow, most finds consist of only skeletal portions, and complete skeletons are harder to find. During molting, a suture opened up across the head, and the trilobite simply fell out of its exoskeleton. But sometimes, a clean suture failed to open, and the animal had to wiggle out

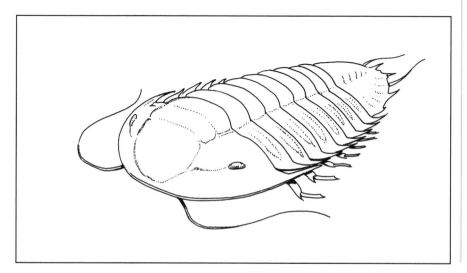

Figure 80 *Trilobites first appeared in abundance in the Cambrian and became extinct at the close of the Paleozoic.*

of its outer covering. In either case, it was still vulnerable to predators until its new skeleton hardened.

Trilobite fossils are often found with their bodies completely curled up as a protection against predators. Many trilobite fossils show rounded bite scars, predominantly on the right side of the body. Predators might have attacked from this direction because when the trilobite curled up to protect itself it exposed its right side. If the trilobite had a vital organ on its left side and was attacked there, it stood a good chance of being eaten, thereby leaving no fossil. But if attacked on the right side, the trilobite had a better chance of entering the fossil record albeit with a chunk bitten out of its body.

Because arthropods must shed their outer skeletons in order to grow, a single individual could leave several fossils behind. Arthropods ranged from the early Cambrian (possibly late Precambrian) to the present.

ECHINODERMS

The echinoderms, from Greek meaning "spiny skin," are unique among the more complex animals. They exhibit both bilateral and fivefold radial symmetry, with arms radiating outward from the center of the body. Echinoderms are also the only animals to possess a system of internal canals, called a water vascular system, that operates a series of tube feet used for locomotion, feeding, and respiration. The great success of the echinoderms in thriving in their environment is illustrated by the existence of more classes of this animal both living and extinct than of any other phylum. The five major classes of living echinoderms are starfish, brittle stars, sea urchins, sea cucumbers, and sea lilies. The fossil record of the echinoderms goes back to the Cambrian and possibly late Precambrian.

The crinoids (Fig. 81), commonly known as sea lilies, became the dominant echinoderms of the middle and upper Paleozoic, and some species still exist. They had long stalks, some more than 10 feet in length, composed of up to 100 or more calcite disks, and were anchored to the ocean floor by a rootlike appendage. Perched atop the stalk was a cup called a calyx that housed the digestive and reproductive systems. Food particles were strained from passing water currents by five feathery arms that extended from the cup, giving the animal a flowerlike appearance. The extinct Paleozoic crinoids and their blastoid relatives, whose calyx resembled a rosebud, made excellent fossils, especially the stalks, which on weathered limestone outcrops look like large strings of beads (Fig. 82).

The starfish are common today and left fossils in Ordovician rocks of the central and eastern United States. They have skeletons composed of tiny plates that are not rigidly joined. As a result, the skeleton usually disintegrated when

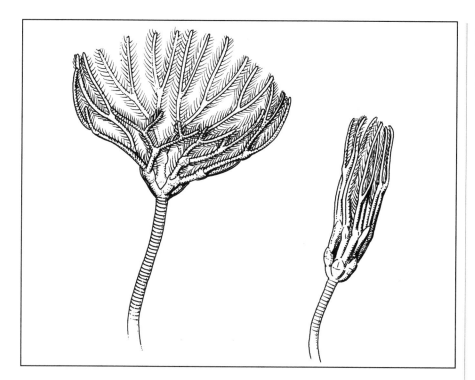

Figure 81 *Crinoids were a dominant species in the middle and late Paleozoic and are still in existence today.*

the animal died, making whole starfish fossils rare. The sea cucumbers, named so because of their shape, have large tube feet modified into tentacles. They have a skeleton composed of isolated plates, which are occasionally found as fossils.

One of the most significant fossil groups is the echinoid, which includes sea urchins, heart urchins, and sand dollars. Their skeletons are composed of limy plates and are characteristically spiny, spherical, or radially symmetrical. Some more advanced forms were elongated and bilaterally symmetrical. The sea urchins lived mostly among rocks encrusted with algae, upon which they fed. Unfortunately, by its very exposure such an environment was not very conducive to fossilization. The familiar sand dollars, which occasionally wash up on beaches, likewise are rare in the fossil record.

From fossil specimens paleontologists have learned of the existence of a number of strange animals that have defied efforts to classify them into existing phyla. One of these animals, appropriately named hallucigenia (Fig. 83), propelled itself across the ocean floor on seven pairs of pointed stilts. Seven tentacles arose from the upper body, and each appeared to contain its own mouth. An alternative interpretation holds that the stilts were actually protective spines along the back, and the tentacles were the true legs. Another odd animal had five eyes arranged across its head, a vertical tail fin to help steer it across the seafloor, and a grasping organ that projected forward for catching prey. One unusual worm had enormous eyes and prominent fins.

Graptolites, colonies of cupped organisms that resembled stems and leaves, look much like plants but were actually animals. They were fixed to the ocean floor like small shrubs, floated freely near the surface, or were attached to seaweed. Large numbers of these organisms were buried in the bottom mud and fossilized into carbon, producing organic-rich black shales. This stratum is so common the world over that graptolites are the most important group of fossils for long-distance time correlation of the lower Paleozoic. Graptolites were thought to have gone extinct in the late Carboniferous about 300 million years ago, but the discovery of living pterobranchs, possible modern counterparts of graptolites, suggests that these might be "living fossils."

Conodonts are tiny jawlike appendages that commonly occurred in Paleozoic marine rocks from 520 million to 205 million years ago and are important markers for dating sediments of this era. They are among the most baffling of all fossils and have puzzled paleontologists since the 1800s, when they began finding these isolated toothlike objects in rocks from the late Cambrian through the Triassic periods. Conodonts show their greatest diver-

Figure 82 *Cogwheel-shaped crinoid columns in a limestone bed of the Drowning Creek Formation, Flemming County, Kentucky.*

(Photo by R. C. McDowell, courtesy USGS)

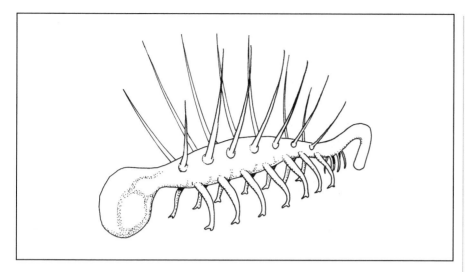

Figure 83 *The Burgess
Shale fauna hallucigenia
is one of the strangest animals preserved in the fossil
record. A previous inter-
pretation had been that it
walked on seven pairs of
stilts and had seven pairs
of tentacles, each with its
own mouth. The current
interpretation holds that
the stiltlike structures were
protective spines and the
tentacles were the legs.*

sity during the Devonian and are important for long-range rock correlations
of that period.

The conodonts appear to be bony appendages of an unusual, soft-
bodied hagfishlike animal that was perhaps the most primitive of vertebrates. But the shape of the missing creature remained unclear until 1983,
when paleontologists discovered toothlike pieces at the forward end of an
eel-shaped fossil from Scotland. Furthermore, the presence of distinctive
eye muscles not known in invertebrates pushed the vertebrate fossil record
as far back as the Cambrian.

VERTEBRATES

The marine vertebrates are well represented by fish, which are a highly successful and diverse group. The most primitive of the chordates, which also include
the vertebrates, was a small, fishlike oddity called amphioxus. Although the animal did not have a backbone, it nonetheless is placed in direct line to the vertebrates. The earliest vertebrates lacked jaws, paired fins on either side of the
body, or true vertebrae and shared many characteristics of modern lampreys. The
origin of the vertebrates led to the evolution of one of the most important nov-
elties—namely, the head. It was packed with paired sensory organs, a complex
three-part brain, and many other features not present in invertebrates.

The vertebrates first appeared in the geologic record about 520 million
years ago. They had internal skeletons made of bone or cartilage, one of life's
most significant advancements. An internal skeleton allowed the wide dispersal of free-swimming species into a variety of environments. The vertebrate

skeleton was light, strong, and flexible with efficient muscle attachments and grew as the animal matured. Invertebrates, supported by external skeletons, were at a distinct disadvantage in terms of mobility and growth. Many animals such as crustaceans had to shed their shells as they grew, thereby becoming vulnerable to predators.

The first vertebrates were probably wormlike creatures with a stiff, spinelike rod called a notochord running down the back to support organs and muscles, a system of nerves along the spine, and rows of muscles arranged in a banded pattern attached to the backbone. Rigid structures made of bone or cuticle acted as levers. Flexible joints efficiently translated muscle contractions into organized movements, such as rapid lateral flicks of the body to propel the animal through the water.

A tail and fins eventually evolved to provide stabilization, and the body became more streamlined and torpedo-shaped for speed. With intense competition among the stationary and slow-moving invertebrates, any advancement in mobility was highly advantageous to the vertebrates. The widespread distribution of primitive fish fossils throughout the world suggests a long vertebrate record early in the Paleozoic.

The first protofish were jawless, generally small (about the size of a minnow), and heavily armored with bony plates. Although they were well protected from their invertebrate enemies, the added weight kept these fish mainly on the bottom, where they sifted mud for food particles and expelled waste through slits on both sides of the throat, which later became gills. Gradually, the protofish acquired jaws with teeth, the bony plates gave way to scales, lateral fins developed on both sides of the lower body for stability, and air bladders supplied buoyancy.

Jaws first developed in fish about 460 million years ago, revolutionizing predation. Giant jawed vertebrates, some of which were monsters in their day, climbed to the very top of the food chain. One of these groups gave rise to land animals, emphasizing the great importance jaws played in vertebrate evolution. The development of jaws also improved fish respiration by supporting the gills. After a fish draws water into its mouth, it squeezes the gill arches to force the water over the gills at the back of the mouth. Blood vessels in the gills exchange oxygen and carbon dioxide as the water flows out the gill slits. The jaws had the advantage of clamping down on very large prey, allowing fish to become fierce predators. Primitive jawed fish might have even caused the demise of the trilobites, once spectacularly successful in the Cambrian seas.

The Devonian period has been popularly called the "age of the fish," and the fossil record reveals so many and varied kinds that paleontologists have a difficult time classifying them all. Fish make up over half the species of vertebrates, both living and extinct. They include the jawless fish—lampreys and hagfish; the cartilaginous fish—sharks, skates, rays, and ratfish; and the bony fish—salmon, swordfish, pickerel, and bass. All major classes of fish alive today had ancestors in the Devonian, but not all Devonian fish species survived to the present. The extinct placoderms were fearsome giants, reaching 30 feet and more in length.

The coelacanths (Fig. 84), which are called "living fossils," are an exception to this rule. Thought to be extinct for 80 million years, six-foot coelacanth was caught in deep water off the coast of Cape Town, South Africa, in 1938. The fish looked ancient, a sort of castaway from the past, with a fleshy tail; a large set of forward fins behind the gills; powerful square, toothy jaws; and heavily armored scales. The most remarkable aspect of the catch was that this fish was not significantly different from fossils of its 460-million-year-old ancestors.

The sharks were highly successful from the Devonian to the present. Instead of skeletons made of bone, as with other fish, sharks utilize cartilage, a more elastic and lighter material. Cartilage does not fossilize well, however, and as a result about the only common remains of ancient sharks are their teeth, which can be found in relative abundance in rocks of Devonian age onward (Fig. 85). Closely related to the sharks are the rays, which are substantially flattened with pectoral fins enlarged into wings up to 20 feet across and a tail reduced to a thin, whiplike appendage.

Establishing connecting links between fish and terrestrial vertebrates were the Devonian crossopterygians and the lungfish, which is another living fossil. The crossopterygians (Fig. 86) were lobe-finned, meaning that the bones in their fins were attached to the skeleton and arranged into primitive elements of a walking limb. They breathed by taking air into primitive nostrils and lungs as well as by using gills, thereby placing them in the direct line of evolution from fish to land-living vertebrates that gave rise to amphibians and reptiles.

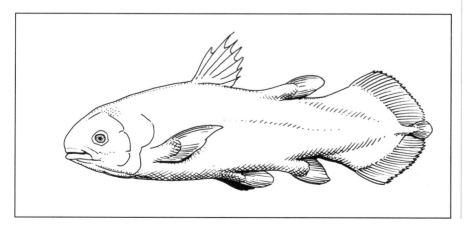

Figure 84 Modern coelacanths have not changed significantly from their 460-million-year-old ancient ancestors.

Reptiles that returned to the sea included the Mesozoic sea serpent–like plesiosaurs, the seacowlike placodonts, and the dolphinlike mixosaurs. The 150-million-year-old pachycostasaurs were carnivorous marine reptiles with thick, heavy ribs possibly used for housing of very large lungs and for ballast in hunting at great depths. The ichthyosaurs were fast-swimming, shell-crushing predators of ammonites that might have been responsible in part for their extinction at the end of the Cretaceous. The placodonts were a group of short, stout marine reptiles with large, flattened teeth and probably fed primarily on bivalves and other mollusks. Several other reptilian species also went to sea, including lizards and turtles that were quite primitive. However, only the smallest turtles survived extinction and lived on to the present.

Cetaceans are marine mammals that include whales, porpoises, and dolphins, all of which evolved during the middle Cenozoic. The dolphins had reached the level of intelligence comparable to that of living species by about 20 million years ago probably because of the stability of their ocean environment. Sea otters, seals, walruses, and manatees are not fully adapted to a continuous life at sea and have retained many of their terrestrial characteristics.

Whales adapted to swimming, diving, and feeding at least as well as fish and sharks. They might have gone through a seallike amphibious stage early in their evolution. Today, their closest relatives are the artiodactyls, or hoofed mammals that have an even number of toes, such as cows, pigs, deer, camels, and giraffes. The ancestor of modern whales apparently was a four-legged car-

Figure 85 *Excavation for shark teeth at Shark's Tooth Hill, Kern County, California.*

(Photo by R. W. Pack, courtesy USGS)

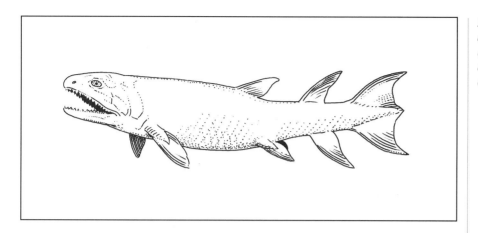

Figure 86 *The crossopterygians were in direct line of evolution to air-breathing, land-dwelling amphibians.*

nivorous mammal that walked on land and swam in rivers and lakes about 57 million years ago. The first whales probably lived in freshwater before entering the sea and did not stray far from the coastline because they needed to return to a river to drink. Ancestors of the giant blue whale, the largest animal on Earth, even dwarfing the biggest dinosaurs that ever lived, evolved from ancient toothed whales about 40 million years ago.

The pinnipeds, meaning "fin-footed," are a group of marine mammals with four flippers, whose three surviving forms include seals, sea lions, and walruses. The "true" seals without ears are thought to have evolved from weasel- or otterlike forms, whereas sea lions and walruses are believed to have developed from bearlike forms. The similarity in their flippers, however, suggests that all pinnipeds evolved from a single land-based mammal that entered the sea millions of years ago.

After learning about fossils of marine species, the next chapter will continue the discussion on fossils of land species.

6

TERRESTRIAL FOSSILS
CREATURES PRESERVED FROM THE LAND

L ike the animals, plants do not appear in the fossil record as complex organisms until the Cambrian, after which they began to evolve rapidly as they entered a new environment. The early seaweeds were soft and easily destroyed upon death and therefore did not fossilize well. Nevertheless, the Cambrian has been called the "age of the seaweed," even though this contention is not strongly supported by fossil evidence. A variety of fossil spores used for reproduction have been found in late Precambrian and Cambrian sediments, findings that suggest that complex sea plants were in existence, but no other significant remains have been found. Even as late as the Ordovician, plant fossils appeared to be composed almost entirely of algae, which probably formed algal mats similar to those on seashores today. Once life crept ashore, however, plants quickly covered the entire Earth's surface with lush forests.

Strangely, life had been in existence for over three-quarters of the Earth's history before it conquered the land some 450 million years ago. Part of the reason might be that the level of oxygen during this time was not sufficient to form an effective ozone screen. The ozone layer, which lies between 25 and 30 miles above the Earth, filters out harmful ultraviolet rays from the

124

Sun. The strong ultraviolet radiation probably kept life in the protective waters of the oceans until such a time as the concentration of ozone became high enough to make conditions safe for venturing onto dry land. This has significant implications for us today, for if we destroy the ozone layer by pumping pollution into the atmosphere, the land could once again be barren of life.

LAND PLANTS

Although terrestrial fossils are much less abundant than marine fossils, primarily because the land is subjected to erosion, some environments such as swamps and marshes have provided an abundance of plant fossils (Fig. 87). With the exclusion of simple algae and bacteria, plants are generally divided into two major categories. The first is the subkingdom of Bryophyta, which includes mosses and liverworts. The second is the subkingdom of Tracheophyta, which includes the higher plants with roots, stems, leaves, and flowers.

Prior to the invasion of terrestrial plants, a slimy coating of photosynthesizing cyanobacteria or blue-green algae might have inhabited the land. For several hundred million years, simple plants had been turning the Earth into

Figure 87 Fossilized plants of the upper Potsville series, Washington County, Arkansas.

(Photo by E. B. Hardin, courtesy USGS)

a more hospitable place for more advanced vegetation. Before the emergence of land plants, microbial soils facilitated the transition to a life out of water. The microorganisms probably formed a dark, knobby soil, resembling lumpy mounds of brown sugar spread over the landscape.

The first land plants consisted of algae and early seaweeds, residing just below the surface in the shallow waters of intertidal zones. Primitive forms of lichen and moss lived on exposed surfaces. They were followed by tiny fern-like plants called psilophytes, or whisk ferns, the predecessors of trees. These simple plants, which lived semisubmerged in the intertidal zones, lacked root systems and leaves and reproduced by casting spores attached to the ends of simple limbs into the sea for dispersal. The most complex land plants grew less than an inch tall and resembled an outdoor carpet covering the landscape.

The most significant evolutionary step was the development of a vascular stem to conduct water to the plant's extremities, which enabled plants to live inland successfully. The early club mosses, ferns, and horsetails were the first plants to utilize this water vascular system. The lycopods, which include the club mosses and scale trees, were the first to develop true roots and leaves. The leaves were generally small and the branches were arranged in a spiral. The spores were attached to modified leaves that became primitive cones. The scale trees (Fig. 88), so named because scars on their trunks resembled large fish scales, grew up to 100 feet tall and were among the dominant trees of the Pennsylvanian coal swamps.

As leaves became larger, branches had to be strengthened so they would not break as the tree continued to grow. In order to maximize photosynthesis, the creation of carbohydrates by the reaction of sunlight, water, and carbon dioxide, leaves were exposed to as much sunlight as possible, placing further mechanical stress on the plant especially during windstorms. Therefore, those plants that developed an efficient branching pattern that gathered most light were the most successful. The best branching pattern comprised tiers of branches similar to those of present-day pines. This pattern emerged during the first 50 million years of land plant life and remains highly successful today.

The mountainous landscape in the northern latitudes of the Paleozoic supercontinent Pangaea were dominated by thick forests of primitive conifers, horsetails, and club mosses that grew to 30 feet tall. Much of the continental interior probably resembled a grassless rendition of the contemporary steppes of central Asia, with temperatures varying from very hot in summer to extremely cold in winter. Since grasses were not yet in existence, the scrubby landscape was dotted with bamboolike horsetails and bushy clumps of now-extinct seed ferns resembling present-day tree ferns.

When the Tropics grew more arid and the Carboniferous swamplands disappeared, the climate change initiated a wave of extinctions that wiped out practically all lycopods at the beginning of the Permian 280 million years ago. Today, they only exist as small grasslike plants in the Tropics. As the climate

grew wetter and the swamps reemerged, weedy plants called tree ferns dominated the Paleozoic wetlands.

The second most diverse group of living plants are the true ferns. They ranged from the Devonian to the present and were particularly widespread in the Mesozoic, when they prospered in the mild climates, whereas today, they are restricted to the Tropics. Some ancient ferns attained heights of present-day trees. The Permian seed fern glossopteris was especially significant for providing further evidence for continental drift. Its fossil leaves (Fig. 89) were abundant on the southern continent Gondwana but were suspiciously missing on the northern continent Laurasia, indicating that the two megacontinents were far apart and divided by the great Tethys Sea.

The gymnosperms, including cycads, ginkgos, and conifers, originated in the Permian and bore seeds lacking fruit coverings. The cycads, which resembled palm trees, were highly successful during the Mesozoic and ranged across all major continents. They probably contributed significantly to the diets of the plant-eating dinosaurs. The ginkgo, of which the maidenhair tree in east-

Figure 89 *The existence of fossil glossopteris leaves helped prove the theory of continental drift.*

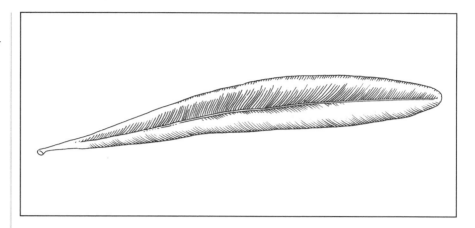

ern China is the only living relative, might be the oldest genus of seed plant. Also dominating the Mesozoic forests were the conifers, some of which reached nearly 400 feet high. Petrified trunks of conifers (Fig. 90) were found to be as much as five feet in diameter and 100 feet tall.

By far the largest group of plants in the world today are the angiosperms, or flowering plants. The sudden appearance of angiosperms and their eventual domination of plant life have remained a mystery, however. They were dis-

Figure 90 *The three most prominent petrified tree stumps on North Scarp, Specimen Ridge, Yellowstone National Park, Wyoming.*

(Courtesy National Park Service)

tributed worldwide by the end of the Cretaceous, and today they include about 270,000 species of trees, shrubs, grasses, and herbs. Flowering plants evolved alongside pollinating insects and offered them brightly colored and scented flowers along with sweet nectar. Many angiosperms also depended on animals to disperse their seeds, which were encased in fruit with an attractive taste. The seeds passed through the animal's digestive tract and were dropped some distance away.

The earliest angiosperms apparently were large plants as tall as magnolia trees. Yet fossils discovered in Australia suggest that the first angiosperms there and perhaps elsewhere were merely small herblike plants. Within a few million years after they burst onto the evolutionary scene, the efficient flowering plants crowded out the once-dominant gymnosperms and ferns. They contained water-conducting cells called vessel elements that enabled the advanced plants to cope with drought conditions. Before such vessels appeared, plants were restricted to moist areas such as the wet undergrowth of rain forests. The angiosperms were widely distributed by the end of the Cretaceous.

The major groups of modern plants were represented in the early Tertiary period. The angiosperms dominated the plant world; all modern families had evolved by about 25 million years ago. Grasses were the most significant angiosperms, providing food for hoofed mammals called ungulates. Their grazing habits evolved in response to the widespread proliferation of grasslands, which sparked the evolution of large herbivorous mammals and in turn ferocious carnivores to prey on them.

AMPHIBIANS

The first animals to crawl out of the sea were probably crustaceans, segmented creatures, ancestors of today's millipedes, that walked on perhaps 100 pairs of legs. The earliest terrestrial communities supported small plant-eating arthropods, which served as prey for advanced predatory animals when they finally went ashore. They remained near the shore at first and later moved farther inland along with the early mosses and lichens. With the land to themselves and lacking competition or predation, these creatures found plenty to feed upon, and some grew to giants, six feet long. However, when relatives of the giant sea scorpion eurypterid took to the land, the ancestral millipedes, which were too slow to escape them, became easy prey.

With the advent of the forests, leaves and other edible parts were no longer near the ground within easy reach, posing serious problems for the ancestors of the insects. By developing winglike appendages, which might have been originally intended as cooling devices, insects were able to launch themselves into the forest canopy. The adaptation was particularly useful for escaping hungry amphibians, which were beginning to appear about this time.

Prior to the amphibian invasion, freshwater invertebrates and fish had been inhabiting lakes and streams. The vertebrates had spent more than 160 million years underwater, with only a few short forays onto the land. Relatives of the lungfish lived in freshwater pools that dried out during seasonal droughts, requiring the fish to breathe with primitive lungs as they crawled to the safety of the nearest water hole (Fig. 91).

The amphibious fish probably spent a limited time on shore because their primitive legs could only support their body weight for short periods, requiring them to return to sea. As their limbs strengthened, the amphibious fish wandered farther inland within accessible distances of nearby water sources such as swamps or streams where crustaceans and insects were abundant. The amphibious fish eventually evolved into the earliest amphibians, and their legacy is well documented in the fossil record.

The apparent ancestors of the amphibians were the crossopterygians, which were encouraged to make short forays on shore to feed on abundant crustaceans and insects during the middle Devonian, about 380 million years ago. They began to strengthen their lobe fins into walking limbs by digging in the sand for food and shelter. This allowed them to venture farther inland, although not too far from accessible swamps or streams. By the early Mississippian, about 335 million years ago, these fish gave rise to the earliest amphibians (Fig. 92). The amphibians still depended on having a nearby source of water to moisten their skins and to allow reproduction, since, like fish, they laid their eggs in water.

The earliest known tetrapod (four-legged animal), called acanthostega (Fig. 93), meaning "spine plate," had a salamanderlike body with large eyes on top of a flat head for spotting prey swimming above as it sat buried in the bottom mud. It sported eight toes on each front leg, perhaps the most primitive

Figure 91 *Air-breathing fish traveled overland to new water holes.*

Figure 92 *The rough evolution from the crossopterygian (top) to the amphibious fish (middle) to the amphibian (bottom). Creatures are not shown to scale.*

of walking limbs. The digits were sophisticated and multijointed, but because they were attached to an insubstantial wrist, the legs were virtually useless for walking on the ground. The rest of the skeletal anatomical characteristics also suggest acanthostega could not have walked on land easily and instead probably crawled around on the bottom of lagoons and used gills for respiration.

An ancient land vertebrate called ichthyostega, meaning "fish plate," was one of the earliest known amphibians; it lived half in and half out of water. It was dog sized and had a broad, flat fishlike head and a small fin atop the tail, apparently used for swimming. It developed a sturdy rib cage to hold up its internal organs while on land and crawled around on primitive legs with seven toes on each hind foot. Amphibians also sported six and eight digits on their feet, indicating that the evolution of early land vertebrates followed a flexible pattern of development. But no terrestrial vertebrates evolved a foot with more than five true digits. Neither acanthostega nor ichthyostega could do much more than waddle around on land. Their upper arm bones had a broad, blobby shape ill suited to walking. Their hind limbs splayed out to the side and could not have easily held up the body.

Figure 93 *Acanthostega walked on the bottom of lagoons with legs having eight toes.*

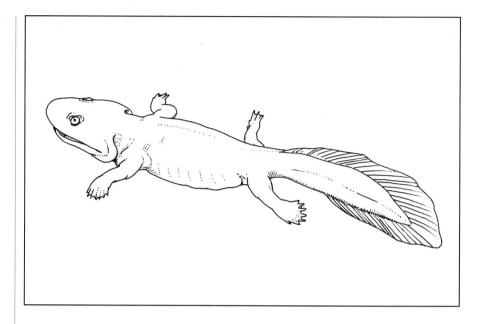

Around 330 million years ago, the tetrapods branched into two groups, one leading to amphibians and the other to reptiles, dinosaurs, birds, and mammals. Some amphibians had strong, toothy jaws and resembled giant salamanders up to five feet in length. A two-foot-long amphibian with armadillolike plates rooted in the soil for worms and snails.

The early amphibians were slow and ungainly creatures, whose weak legs could hardly keep their squat bodies off the ground for any length of time. Thus, in order to become successful hunters despite lack of speed and agility, the amphibians developed quite a remarkable tongue that could lash out at insects and flick them into their mouth. The adaption worked so well that the amphibians quickly populated the Earth. But the necessity of a semi-aquatic lifestyle led to the eventual downfall of the amphibians when the great swamps began to dry up toward the end of the Paleozoic.

The amphibians continued to decline in the Mesozoic with all large flat-headed species becoming extinct early in the Triassic. The group thereafter was represented by the more familiar toads, frogs, and salamanders. The fossil remains of these amphibians are largely fragmentary because vertebrate skeletons are constructed with a large number of bones that are easily scattered by surface erosion.

REPTILES

The vacancy left behind by the amphibians was filled by their cousins, the reptiles, which were better suited for a full-time life out of water. One of the reasons for the reptiles' great success was their more efficient means of locomotion. The improvement over the amphibian foot included changes in the form of the toes; the addition of a short, thumblike fifth toe; and the appearance of claws. In addition, the reptiles' toes pointed forward to allow the animals to run, whereas the amphibians' toes were splayed outward, making movement slow and ungainly.

Reptiles have scales that retain the animals' bodily fluids, whereas amphibians have a permeable skin that helps them breathe. The skin of the amphibian also had to be moistened, and the animal would dry up if away from water for long periods. Reptiles lay eggs with hard watertight shells on dry land, whereas the amphibians' eggs do not have a protective membrane and as with fish their eggs had to be laid in water or moist places. Because the embryos took more time to develop, reptile eggs had to be protected from predators. This parental attention gave the young a better chance of survival, which contributed significantly to the reptile's great success in populating the land.

Like fish and amphibians, reptiles are cold-blooded; that does not necessarily mean that their blood is always cold. Instead, the temperature of their bodies depends on the temperature of the environment because reptiles have no means of regulating their body temperatures. Therefore, they are rather sluggish on cold mornings and must wait until the sun warms their bodies before reaching their peak performance. The unusually warm climate of the Mesozoic must have contributed substantially to the success of the reptiles. Furthermore, reptiles only required a fraction of the amount of food that mammals the same size need to survive because mammals use most of their calories to maintain their high body temperature.

The reptiles became the leading form of animal life on Earth and occupied land, sea, and air. The seacowlike placodonts, the sea serpent–like plesiosaurs, and the dolphinlike ichthyosaurs (Fig. 94) were reptiles that returned to the sea to compete with fish and achieved great success. Lizards and turtles

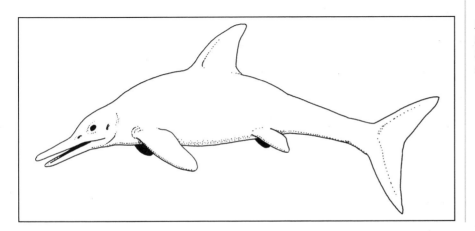

Figure 94 The ichthyosaur was an air-breathing reptile that returned to the sea.

Figure 95 *The giant pterosaur ruled the skies for 120 million years.*

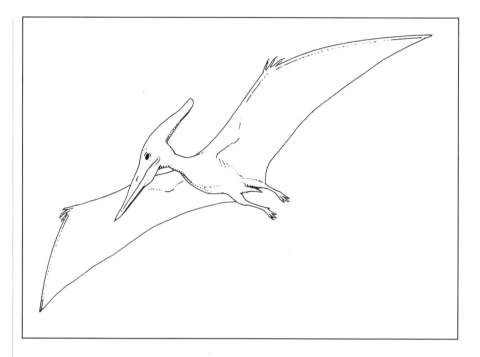

also went to sea, and many modern giant turtles are descendants of those marine reptiles.

Some early reptiles were large animals. Moschops were up to 16 feet long and had thick skulls adapted for butting during mating season, much like the behavior of modern herd animals. They might have been preyed upon by packs of lycaenops, which were reptiles with doglike bodies and long canine teeth projecting from the mouth. Phytosaurs were large, heavily armored predatory reptiles sporting sharp teeth. They resembled crocodiles, with short legs, long tails, and elongated snouts, but were not closely related to them. They evolved from the thecodonts, the same group that gave rise to crocodiles and dinosaurs. The phytosaurs thrived in the late Triassic, evolving quite rapidly, but apparently did not survive beyond the end of the period.

Perhaps the most spectacular reptiles that ever existed were the flying pterosaurs (Fig. 95). They had wingspans of up to 40 feet, about the size of a small aircraft, and dominated the skies for 120 million years. Their wings were similar in construction to bats' wings with a finger on each forelimb greatly elongated and covered with a membrane attached along the side of the body. This appendage probably originated as a cooling mechanism, and the reptile regulated its body temperature by fanning its forelimbs. When the pterosaur wanted to fly, it simply caught a breeze, and with a single flap of its enormous wings and a kick from its powerful hind legs, it was airborne. It probably spent most of its time aloft riding updrafts as modern sailplanes do.

By the close of the Triassic, a remarkable reptilian group appeared in the fossil record. These were the alligators and crocodiles. Members of this group adapted to life on dry land, a semiaquatic life, or a fully aquatic life. One marine species was about 15 feet in length and had a streamlined head, a sharklike tail, and legs molded into swimming paddles. Fossil crocodiles were among the first vertebrates unearthed by early 19th-century paleontologists and were used to support Charles Darwin's theory of evolution. Fossil alligators and crocodiles are found in the higher latitudes as far north as Labrador, indicating an unusually warm climate during the Cretaceous (Fig. 96).

DINOSAURS

Dinosaurs descended from the thecodonts, the common ancestors of crocodiles and birds, making them distant living relatives of the dinosaurs. One group of thecodonts returned to the sea and became large fish-eating aquatic species. They included the phytosaurs, which died out in the Triassic, and the crocodilians, which remain successful today. Pterosaurs were also descendants of the thecodonts. The appearance of featherlike scales ostensibly used for insulation suggests that thecodonts were also the ancestors of birds. The

Figure 96 *The Chinle Formation, Uinta Range, Utah, where crocodile fossils are found, indicating an extremely warm climate during the Cretaceous period.*

protofeathers helped trap body heat or served as a colorful display for attracting mates, as in modern birds.

The dinosaurs are classified into two major groups: the sauropods, which were mostly herbivores, and the carnosars, which were mostly carnivores. Not all dinosaurs were giants, however, and many were no larger than most modern mammals. The smallest known dinosaur footprints are only about the size of a penny. The smaller dinosaurs had hollow bones similar to those of birds. Some had long, slender hind legs; long, delicate forelimbs; and a long neck, and if not for a long tail their skeletons would have resembled those of ostriches.

Many of the smaller dinosaurs established a permanent two-legged stance, which increased their speed and agility and freed their forelimbs for more successful foraging. The large, two-legged carnivores were swift runners and highly aggressive. Tyrannosaurus rex (Fig. 97), the greatest land carnivore that ever lived, stood on powerful hind legs with its forelimbs shortened to almost useless appendages. The animals probably traveled in packs and hunted huge herds of hadrasaurs as a pride of African lions hunts gazelle today.

Figure 97 *Tyrannosaurus rex was the greatest land carnivore that ever lived.*

The carnivorous dinosaurs were cunning and aggressive, attacking their prey with unusual voracity. The cranial capacity of some carnivores suggests they had relatively large brains and were fairly intelligent, able to react to a variety of environmental pressures. The velociraptors, meaning "speedy hunters," with their sharp claws and powerful jaws were vicious killing machines, whose voracious appetites suggest they were warm-blooded. They are also considered close relatives of birds.

A great debate rages today among paleontologists over whether dinosaurs are cold-blooded as reptiles or are warm-blooded as are mammals and birds. One argument in favor of warm-bloodedness contends that the skeletons of the smaller, lighter dinosaurs bear a resemblance to those of birds, which are themselves warm-blooded. Evidence for rapid juvenile growth, which is common among mammals, is found in the bones of some dinosaur species, possibly providing another sign of warm-bloodedness. The skulls of some dinosaurs show evidence of respiratory turbinates, which act as heat exchangers in the nostrils of warm-blooded animals.

Some dinosaurs are thought to have been swift and agile, requiring a high rate of metabolism that only a warm-blooded body could provide. The complex social behavior of the dinosaurs also might be an evolutionary advancement that resulted from warm-bloodedness. Perhaps the females of some species gave live births. The major problem with the theory of warm-bloodedness, however, is that at the end of the Cretaceous, when the climate supposedly grew colder, the warm-blooded mammals survived whereas the dinosaurs did not.

During the Jurassic, dinosaurs attained their largest sizes and longest life spans. The biggest dinosaurs occupied Gondwana, which included all the southern continents. The generally warm climate and excellent growing conditions encouraged the growth of lush vegetation, including ferns and cycads, to satisfy the insatiable appetites of the plant-eating dinosaurs. The huge size of the herbivores spurred the evolution of giant carnivorous dinosaurs to prey on them.

Among the largest dinosaur species were the apathosaurs and brachiosaurs, which lived up to about 100 million years ago. These gargantuan creatures fully deserve the title "thunder lizards." They were sauropods with long, slender necks and tails, and the front legs were generally longer than the hind legs. Perhaps the tallest and heaviest dinosaur thus far discovered was the 80-ton ultrasaurus, which could tower above a five-story building. Seismosaurus, meaning "earth shaker," was the longest known dinosaur, reaching a length of more than 140 feet from its head, supported by a long, slender neck, to the tip of its even longer whiplike tail. Giganotosaurus even challenges Tyrannosaurus rex as the most ferocious terrestrial carnivore ever to have lived.

Some dinosaur species grew to such giants for probably the same reasons that the elephant and rhinoceros are big. The majority of the large dinosaurs were herbivores, or plant eaters, and therefore had to consume huge quantities of coarse cellulose, which took a long time to digest. This required an oversized stomach for the fermentation process and, consequently, a large body to carry it around. Some species swallowed gizzard stones (as modern birds do) in order to grind the fibrous fronds into pulp. The rounded, polished stones were left in a heap where the dinosaur died, and deposits of these stones can be found on exposed Mesozoic sediments, especially in the American West.

Because large reptiles possess a capacity for almost unlimited growth, they continue to grow throughout their lives. A large body helps cold-blooded animals maintain their body temperature for long periods. This makes the animal less susceptible to short-term temperature variations in the environment. The only factor that prevent the dinosaurs from growing larger than they did was the force of gravity. When an animal doubles its size, the weight on its bones quadruples. The only exception were the dinosaurs that lived permanently in the sea, and as with present-day whales, some of which are larger than the largest dinosaur that ever lived, the buoyancy of seawater kept the weight off their bones.

Oviraptor, whose name means "egg hunter," was originally thought to have raided nests of other dinosaurs. But a fossil oviraptor was found in the Gobi Desert sitting on a nest filled with as many as two dozen eggs neatly laid out in a circle. The dinosaur resembled a wingless version of an ostrich with a shortened neck and a long tail. It sat with its pelvis in the middle of the nest and had its long arms wrapped around the nest in the same way birds do, perhaps protecting the eggs against a gigantic sandstorm that apparently engulfed and fossilized it along with its clutch 70 to 80 million years ago.

Prior to the discovery, nests with eggs and whole infant dinosaurs had been found but never direct evidence of parents' squatting on eggs. The oviraptor was in the exact position a chicken would take sitting on a nest. Whether the dinosaur was keeping the eggs warm as birds do, shading them from the hot sun, or protecting them is unknown. The clutch was probably a communal nest similar to that used by ostriches, into which hens deposit their eggs and take turns incubating them. The discovery provided the strongest evidence of parental attention, suggesting birds inherited this behavior from dinosaurs.

One reason the dinosaurs were so successful is that they might have nurtured and fiercely protected their young, thus allowing a greater number to mature into adulthood. The parents might have taken food to their offspring, regurgitating seeds and berries similarly to the way modern birds do. Some giant herbivores might have traveled in large herds with the juveniles kept in the middle for protection. Fifteen-foot-tall duck-billed hadrasaurs (Fig. 98) lived in the arctic regions of the Northern and Southern Hemispheres and

Figure 98 *Fossil bones of hadrasaur found in Alaska suggest the dinosaur might have migrated south when the climate grew cold.*

either adapted to the cold and the dark or migrated long distances to warmer regions. Dinosaurs are also thought to have been fairly intelligent and able to react to environmental pressures, capacities that explain why they dominated the planet for as long as they did.

The success of the dinosaurs is exemplified by their extensive range, in which they occupied a wide variety of ecologic niches and dominated all other forms of land animals. Roughly 500 dinosaur genera have been identified, although this number is probably far short of the total. Dinosaurs are known to have ventured to all major continents, and their distribution is strong evidence for continental drift. After the continents rifted apart in the Jurassic, the dinosaurs could no longer cross over from one continent to another, except possibly on a few land bridges, and certain species are no longer found in some areas. Continental drift also might have contributed to the demise of the dinosaurs by rearranging the ocean basins, thus redirecting ocean currents, thereby changing global atmospheric patterns and bringing on unstable climatic conditions.

At the end of the Cretaceous, the dinosaurs along with 70 percent of all known species became extinct. Because the boundary between the Cretaceous and the Tertiary, called the K–T boundary, is not a sharp break, but might represent up to a million years or more, this extinction was not necessarily sudden but could have taken place over an extended period. Many dinosaurs along with other species were already in decline several million years prior to the end of the Cretaceous. Triceratops (Fig. 99), whose vast herds covered the entire globe and might have contributed to the decline of other dinosaur

Figure 99 *Vast herds of triceratops roamed all parts of the world toward the end of the Cretaceous period.*

species by destroying huge habitat areas or spreading disease, were among the last dinosaurs to go.

BIRDS

Birds appeared in the Jurassic, around 150 million years ago, although some accounts place their origin as much as 75 million years earlier. By about 135 million years ago, early birds began to diversify, diverging into two lineages, one leading to modern birds, the other to archaic birds. Birds descended from the thecodonts that gave rise to the crocodilians and the dinosaurs, and for this reason birds have been called "glorified reptiles."

The earliest known fossil bird, called Archaeopteryx (Fig. 100), from Greek meaning "ancient wing," was about the size of a modern pigeon and appeared to be a species in transition between reptiles and true birds. It was thought to be a small dinosaur until fossils clearly showing impressions of feathers were discovered in 1863 in a unique limestone formation in Bavaria, Germany. But unlike modern birds, Archaeopteryx lacked a keeled sternum for the attachment of flight muscles. It also had teeth, claws, a long bony tail, and many skeletal features of small dinosaurs but lacked hollow bones to conserve weight. Its feathers were outgrowths of scales that were probably originally used for insulation.

Although Archaeopteryx appeared to have the necessary accoutrements for flight, it likely was a poor flyer, taking off only for short distances as do today's domesticated birds. It probably became airborne by running along the ground with its wings outstretched and gliding for a brief moment or by leaping upward while flapping its wings to catch insects flying by. Its forebears might have flapped their wings in order to increase running speed while escaping predators, thereby obtaining flight by pure accident.

Birds are warm-blooded in order to obtain the maximum sustainable amount of energy needed for flight, which requires a high metabolic rate. They also retain the reptilian mode of reproduction by laying eggs. The bones of some Cretaceous birds show growth rings, a feature common among cold-blooded reptiles, suggesting that early birds might not have yet developed fully warm-blooded bodies. The bird's ability to maintain high body temperatures has sparked a controversy over whether some dinosaur species with similar skeletons were warm-blooded as well.

Many bird species retained their teeth until the late Cretaceous, roughly 70 million years ago. Claws on the forward edges of the wings might have helped the bird climb trees, from which it could launch itself into the air. After mastering the skill of flight, birds quickly radiated into all environments. Their superior adaptability enabled them to compete successfully with the pterosaurs, possibly leading to the flying reptile's eventual decline.

Giant flightless birds appeared early in the bird fossil record. Their wide distribution is further evidenced for the existence of the supercontinent Pangaea since these birds had to travel on foot to get from one part of the world to another. Having been driven into the air by the dinosaurs, birds, once the dinosaurs disappeared, found life a lot easier on the ground because they had to expend a

Figure 100 *This Archaeopteryx might be a link between reptiles and birds.*

great deal of energy in order to stay in the air. Some birds also successfully adapted to a life in the sea. Penguins are large flightless birds that have taken to life in the ocean and are well adapted to survive in the Antarctic. Certain diving ducks are especially equipped for "flying" underwater to catch fish.

MAMMALS

The family of mammallike reptiles was in transition from reptiles to mammals. Fossilized bones of a mammallike reptile with large down-pointing tusks called lystrosaurus were discovered in the Transantarctic Range of Antarctica, which, unlike today, was largely free of ice. An ancestor to mammals, lystrosaurus was the most common vertebrate on land and was found throughout Pangaea. Mammallike reptiles called dicynodonts also had two caninelike tusks and fed on small animals along riverbanks.

About 300 million years ago, the pelycosaurs became the first group of animals to depart from the basic reptilian stock. The 11-foot-long dimetrodon (Fig. 101) had a large dorsal fin that was well supplied with blood and is believed to be the first attempt at regulating the body's temperature. As blood circulated through the sail, its temperature was lowered by cooler air in the atmosphere around it. When the climate became warmer, the pelycosars lost their sails and perhaps gained some degree of internal temperature control. The pelycosars thrived for about 50 million years and then gave way to the mammallike reptiles called the therapsids.

Figure 101 *This 300-million-year-old pelycosar called dimetrodon used the huge sail on its back to regulate body temperature.*

The first therapsids retained many characteristics of the pelycosaurs, including legs well adapted for fast running. They ranged in size from that of a mouse to that of a hippopotamus. Early members invaded Gondwana during the late Permian, when that continent was still recovering from an ice age, indicating that the animals might have been warm-blooded in order to survive the cold. They therefore seemed to be well adapted for feeding and traveling through the snows of the cold winters. The therapsids were apparently too large to hibernate in winter, as evidenced by the absence of growth rings in their bones, similarly to the way tree rings indicate different growth rates during the seasons.

As the more advanced therapsids moved into the colder regions, they developed fur in place of scales. They retained the reptilian mode of laying eggs; however, they probably incubated their eggs and nurtured their young. This behavior possibly led to longer egg retention and live birth. The therapsids dominated animal life on Earth for more than 40 million years. Then, for unknown reasons, they lost out to the dinosaurs and were relegated to a nocturnal lifestyle until the dinosaurs became extinct.

Mammals descended from the mammallike reptiles, which were eventually driven into extinction by the dinosaurs about 160 million years ago. The early mammals evolved over a period of more than 100 million years into the first therian (live-birth) mammals, the ancestors of all living mammals, including marsupials and placentals. During this time, mammals progressed toward better functioning in a terrestrial environment. Teeth evolved from simple cones that were replaced repeatedly during the animal's lifetime to more complex forms that were replaced only once. However, the mammalian jaw and other parts of the skull still shared many similarities with those of reptiles. One mysterious group known as the triconodonts, ranging from 150 to 80 million years ago, were primitive protomammals, possibly ancestors of the monotremes represented today by the platypus and echidna, which lay eggs and walk with a reptilian sprawling gait.

Distinguishable features of mammals include four-chambered hearts, a single bone in the lower jaw, highly differentiated teeth, and three small ear bones that migrated from the jawbone backward as the brain grew larger, greatly improving the ability to hear. Mammals give live births and possess mammary glands that provide a rich milk to suckle their young, which generally are born helpless. Mammals have the largest brains, capable of storing and retaining impressions. Therefore, they lived by their wits, an ability that explains their great success. They conquered land, sea, and air and are established, if only seasonally, in all parts of the world.

Of the nearly two dozen orders of mammals that existed after the extinction of the dinosaurs, only half were found in the Cretaceous, and 18 mammalian orders survive today. Many of the archaic mammals, including some large, peculiar-looking animals, disappeared at the end of the Eocene, about 37 million years ago, when the planet took a plunge into a colder climate. Afterward, truly modern mammals began to appear (Fig. 102). The extremes in climate and geographic characteristics during the Cenozoic produced a large variety of living conditions and presented many challenging opportunities for the mammals.

Continental drift isolated many groups of mammals, and they evolved along independent lines. Australia was inhabited by strange, egg-laying mammals called monotremes, which included the spiny anteater and the platypus, both of which rightfully should be classified as surviving mammallike reptiles. Appearing to have been designed by a committee, the platypus has a ducklike bill, webbed feet, and a broad flattened tail and lays eggs. The marsupials have pouches on their bellies for incubating their tiny young after birth. Their ancestors originated in North America about 100 million years ago and migrated southward to Australia, using Antarctica as a land bridge (Fig. 103). Camels also originated in North America in the early Miocene, and 2 million years ago, they migrated to other parts of the world by connecting land bridges, which were exposed during the Pleistocene ice ages.

Horses originated in western North America during the Eocene, when they were only about the size of a small dog. As time went on, they became progressively larger, their faces and teeth grew longer as the animal switched

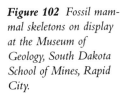

Figure 102 *Fossil mammal skeletons on display at the Museum of Geology, South Dakota School of Mines, Rapid City.*

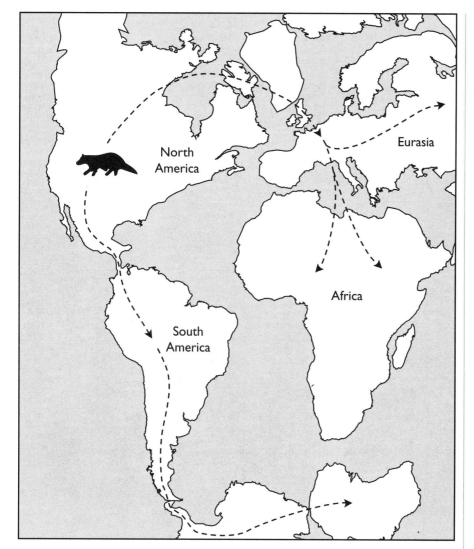

Figure 103 *A map indicating the dispersion of marsupials from North America to other parts of the world 80 million years ago.*

from browsing to grazing, and the toes fused into hoofs. In Africa, the giraffes grew long necks in order to browse on high branches, and the snouts of the elephants were elongated for similar reasons. These are cited as classic examples of evolution at work to adapt animals to their changing environment.

During the last ice age, relatives of the elephant, the mammoth and mastodon, along with other mammals, such as the giant sloth and saber-tooth cat, grew to enormous size and roamed the ice-free regions in many parts of the Northern Hemisphere. The giantism might have resulted from circumstances similar to those that led to the great size of some dinosaurs, including an abundant food supply and lack of predation. When the glaciers retreated, the change to a more seasonal climate broke up uniform environments, causing the forests to shrink and the grasslands to expand, possibly disrupting the food chains of several large mammals. Another theory holds that by this time, humans had become successful hunters and decimated the slow, lumbering creatures, often wastefully, leaving their bones in great heaps (Fig. 104).

The preceding chapters discussed the importance of fossils in Earth history. The following chapters will treat minerals and their importance to geology.

Figure 104 *Fossil mammoth bones on display at the Mammoth Site, Hot Springs, South Dakota.*

7

CRYSTALS AND MINERALS
BASIC BUILDING BLOCKS OF ROCKS

This chapter turns from a discussion of fossils to an examination of minerals. A mineral is a homogeneous substance, with a unique chemical composition and crystal structure. A crystal is an orderly growth of a mineral into a solid geometric form. Most minerals develop crystals, which greatly aid in their identification. The most abundant rock-forming minerals are quartz and feldspar, which make up the majority of the noncarbonate, or crystalline, rocks. When a magma body cools, a variety of minerals with varying crystal sizes separate out of the melt. This process leaves behind highly volatile mineralized fluids that penetrate the country rocks, also called host rocks, surrounding the magma chamber to form veins of ore, from which the mineral can be extracted.

Some of the larger, heavier crystals might sink to the bottom of the magma chamber to form very coarse-grained granitic rocks called pegmatites, from Greek, meaning "fastened together." Minerals are also formed by hydrothermal (hot water) activity, especially on the ocean floor. In many parts of the world, pieces of ocean crust that contain minerals have been shoved up onto land by continental collisions. These ophiolites provide important ore deposits (Fig. 105, 106).

Figure 105 *Metal-rich massive sulfide vein deposit in ophiolite.*

(Courtesy USGS)

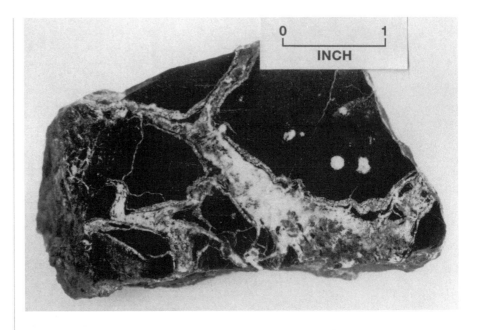

Most minerals consist of two or more elements chemically united in a compound, such as silicon dioxide (SiO_2), which forms quartz. Feldspar, the most common mineral, constituting nearly half the Earth's crust, is composed of aluminum silicates, containing either sodium, calcium, or potassium. Single-element minerals can form metallic ores such as copper or nonmetallic substances such as sulfur, which is mostly associated with volcanic activity (Fig. 107). Graphite, which is the most common form of

Figure 106 *Worldwide distribution of ophiolites, which are slices of oceanic crust shoved onto land by plate tectonics.*

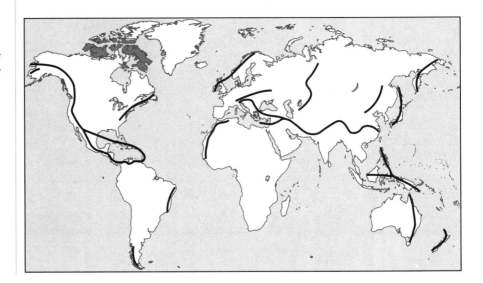

Figure 107 *Sulfur deposits accumulating around a volcanic fumarole.*

(Courtesy USGS)

carbon, has a layered structure, with the chemical bonding within each layer as strong as that in a diamond. Yet the bonding between adjacent layers is so weak that graphite is soft enough to be used for pencil lead or as a lubricant.

CRYSTALS

Crystals are mostly empty space filled with an orderly arrangement of atoms or molecules, and the relative sizes of atoms determine how they fit together to form minerals. In crystals, atoms are arranged in a lattice of identical unit cells, which are the building blocks of the crystal and always contain the same distribution of atoms. Because oxygen exists in the greatest abundance and usually constitutes the largest atoms present, it has the greatest influence on the growth of crystals. Crystals grow by continually adding layers of atoms over a template, or seed crystal, giving each mineral a unique set of crystal planes called faces. For example, large halite crystals (rock salt) can be grown in a container of supersaline sodium chloride solution by the submersion of a single tiny salt crystal.

A crystal's atoms are linked by ionic bonding, whereby positively charged ions called cations are typically surrounded by negatively charged ions

Figure 108 *Large quartz crystals.*

(Photo by W. T. Schaller, courtesy USGS)

called anions. The cluster formed by a cation and its surrounding anions produces regular shapes, with the anions, usually oxygen ions, located on the corners of a polyhedron, such as a tetrahedron with four faces, a cube with six faces, or a octahedron with eight faces. Crystals seem hard because the electrical forces that bind widely scattered atomic nuclei together are extremely powerful. But even sturdy crystals can be compressed and reduced in volume by 50 percent or more when subjected to pressures equivalent to those near the center of the Earth. If the space in which a crystal is growing becomes crowded as a result of the rapid growth of other crystals around it, the crystal could cease growing entirely, or continue growing abnormally. This is the major reason why large, perfect crystals are so rare.

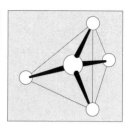

Figure 109 *The basic building blocks of a quartz crystal are tetrahedrons.*

The basic building blocks of a quartz crystal (Fig. 108), the most common oxide of silicon, are tetrahedrons, composed of a silicon ion surrounded by four oxygen ions, situated on the four corners of a tetrahedron (Fig. 109). Each of these shares four corners with four other tetrahedrons to form a continuous three-dimensional framework that is extremely rigid, thus making quartz among the hardest of the common minerals.

In a two-dimensional model of a crystal, an atom sits in the center of a hexagon, whose sides are formed by six of its closest neighbors. This hexagon is the unit cell of the crystal, and crystals can be broken down into a repeating pattern of hexagons. The hexagons in one part of the crystal have the exact same orientation, or alignment, as those in other parts of the crystal. In addition, if straight, parallel lines are drawn connecting all atoms in a crystal, the lines will be evenly spaced across the crystal. Lines from one part of the crystal will match up precisely with lines in another part of the crystal. Many families of these parallel lines exist, each family facing a different direction. In a three-dimensional model of the crystal, these lines become planes called lattice planes.

Every crystal structure has certain symmetries. For example, a crystal has threefold rotational symmetry if the lattice of the crystal looks exactly the same after the crystal is rotated one-third of a circle, or 120 degrees. An example of a shape that has threefold rotational symmetry is an equilateral triangle. Crystals also might have fourfold rotational symmetry such as a square or sixfold rotational symmetry such as a hexagon. But a natural crystal can never have fivefold rotational symmetry because shapes with fivefold rotational symmetry such as pentagons cannot be fitted together without leaving gaps (Fig. 110). In nature, six-sided forms appear in many structures, from honeycombs to lava columns.

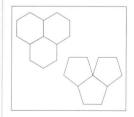

Figure 110 *A comparison of the fit of hexagons and pentagons.*

Crystals are generally classified according to the number, position, and length of their crystal axes (Fig. 111). Isometric crystals like those of halite or galena have three perpendicular axes of equal length. Tetragonal crystals like those of zircon or cassiterite have three perpendicular axes with only two of

151

System	Cubic (isometric)	Tetragonal	Hexagonal	Orthorhombic	Monoclinic	Triclinic
Example	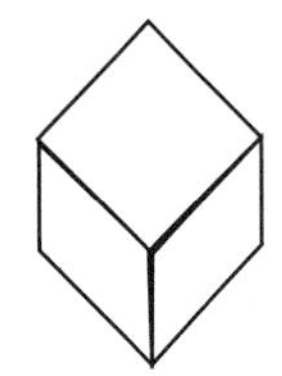		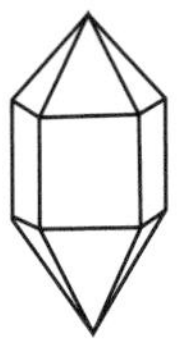	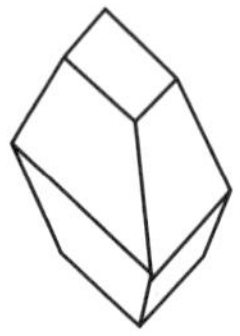	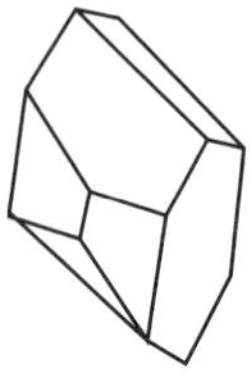	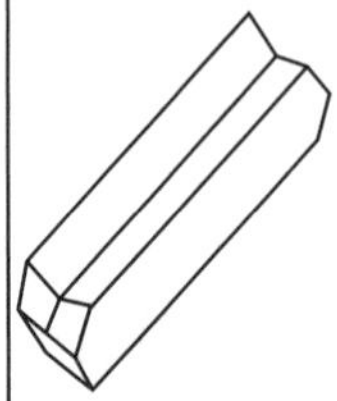
Ideal shape	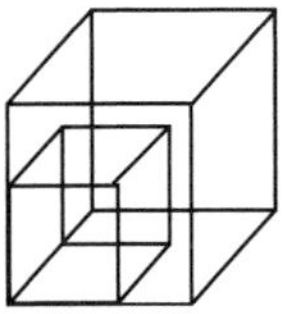	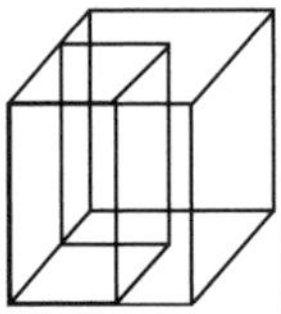	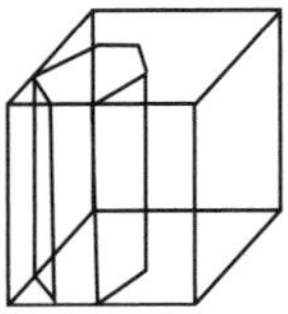	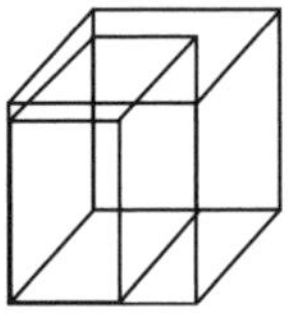	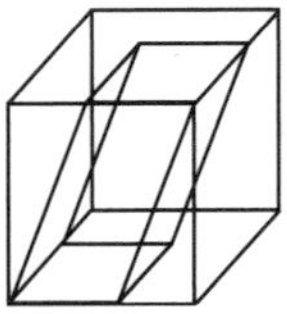	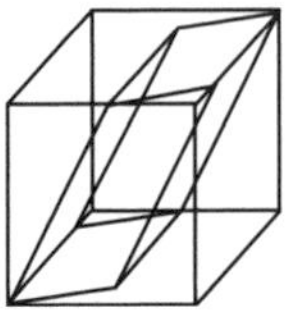
Lengths of axes	 All equal	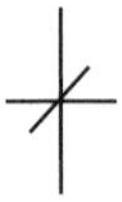 Two horizontal equal–third different	 Three equal horizontal axes—fourth axis different	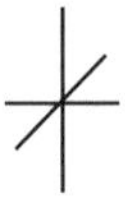 Three axes —unequal	 Three axes —unequal	 Three axes —unequal
Axes of intersection	90°	90°	60° 90°	90°	Only two 90°	None 90°

Figure 111 The classification of crystals depends on their crystal axes, as shown here.

equal length. Hexagonal crystals like those of quartz or calcite have three axes of equal length that meet at angles of 120 degrees and a fourth axis of a different length, perpendicular to the others. Orthorhombic crystals like those of topaz or olivine have three perpendicular axes all with different lengths. Monoclinic crystals like those of gypsum or orthoclase have three unequal axes, two that meet at oblique angles and a third perpendicular to the other two. Triclinic crystals like those of plagioclase and microcline have three unequal axes that meet at oblique angles.

Some minerals can have the same chemical composition but different crystal structures as a result of different environments of deposition. Graphite, one of the softest minerals, and diamond, the hardest natural mineral, are both different forms of the same element, carbon, but are subjected to different environmental conditions such as temperature and pressure. Sometimes, crystals develop in clusters or grow as twins. Twin crystals are two or more crystals of the same mineral that are united with interlocking adjacent parts so that the atomic structure is shared in common. They might grow parallel to each other, grow in opposite directions from each other, or grow in mirror images of each other (Fig. 112). A crystal of one mineral can change by metamorphism into another mineral of the same chemical composition by rearranging its atoms with no change in the crystal's external shape. A crystal of one mineral also can chemically change into another mineral by the addition or loss of elements with no change in the external shape of the crystal.

Figure 112 A twin gypsum crystal showing two halves that are mirror images of each other.

Crystal habit, or appearance, such as cubic, octahedral, or prismatic, is the form a crystal takes in response to temperature, pressure, and other factors in the geologic environment. Some minerals always develop a particular crystalline form; others rarely develop well-formed geometric crystals. Crystals can be prismlike, needlelike, threadlike, bladelike, or sheetlike. They also can be branchlike, netlike, mosslike, or starlike. Crystals can form spherical or semispherical groups, long conic or columnar groups, or concentric groupings of platy, or layered, crystals. They can cover a surface with a layer of closely spaced tiny crystals, called drusy crystals. If crystals grow in restricted spaces, as when magma cools rapidly, the crystal form might only be revealed under a microscope. A few minerals are noncrystalline, lacking an orderly arrangement of molecules, and therefore are called amorphous or massive.

IDENTIFICATION OF MINERALS

Minerals are generally identified by crystal form, color, luster, hardness, density, cleavage, and fracture. Some minerals also can be recognized by transparency, tarnish, tenacity, iridescence, effervescence, fluorescence, magnetism, radioactivity, taste, or smell. In addition to these, geologists identify minerals by using elaborate laboratory equipment, including electron microscopes and X-ray diffraction machines, which provide information on crystal structure.

Because most minerals comprise crystals, the three-dimensional shape of a crystal offers one of the most important clues to identification. The outer arrangement of the plane surfaces reflects the inner structure of the crystal, which in turn is controlled by its chemical composition. If grown without obstruction, minerals develop a characteristic crystal form. Under proper growing condition, crystalline minerals develop exceedingly beautiful forms. Some crystals are so magnificent in their beauty and symmetry that they appear to have been cut and polished. Many perfect crystals are of great value.

Most minerals occur in irregular masses of small crystals as a result of restricted growth; therefore, perfect crystals are rare. Giant crystals are rarer still and reflect a deep-seated origin within the crust. One of the largest crystals was discovered at the Etta mine near Keystone, South Dakota. It was a prismatic spodumene crystal, of the pyroxene group, composed of lithium aluminum silicate, measuring 42 feet long, and weighing approximately 90 tons.

In addition to a crystal's form, its color, a result of the absorption and reflection of certain wavelengths of light, is the most obvious characteristic and another essential clue used in identification. The color of a freshly broken surface of some minerals is a reliable indicator to their identification. Some minerals have a consistent color, such as galena (gray), hematite (red), sulfur (yellow), azurite (blue), and malachite (green). Some minerals, such as quartz, have a variety of hues, which are generally controlled by pigments or impurities. Weathering often changes the color of a mineral, and a fresh surface, obtained by chipping away the outer layers, is required to determine its true color.

Sometimes a mineral's true color does not show until it is ground to a powder or scraped across a piece of unglazed porcelain called a streak plate. Therefore, this color is called the mineral's streak, which is an important clue in identification. The streak of nonmetallic minerals is either colorless or very light, whereas metallic minerals often have a dark streak that might differ from the color of the mineral itself. For example, pyrite has a pale brass-yellow color but its streak is black.

Besides its color, a mineral's luster, which is a property determined by the ability of a mineral to reflect, refract (bend), or absorb light falling on its surface, is important for classifying minerals. The terms used for describing a mineral's luster are those in common everyday usage. A metallic luster is produced by most metals. A brilliant luster is like that of a diamond. Minerals with a glassy luster look like glass. A greasy luster looks oily. Minerals that have the look of resin have a resinous luster. A waxy luster has the appearance of wax. If a mineral has the iridescence of a pearl, it has a pearly luster. If a mineral is fibrous, it has a silky luster. A dull or earthy luster is like that of clay.

The hardness of a mineral is its resistance to abrasion or scratching and is an important aid in identifying minerals. The hardness scale, also called the Mohs' scale, named for the German mineralogist Friedrich Mohs, who first proposed it in 1839, is an arrangement of 10 minerals according to their hardness (Table 9). The softest mineral is talc, which has been used for centuries as a lubricant and as talcum powder. The differences in hardness between corundum and diamond are quite large. However, the hardness of the others is fairly equally spaced along the scale. With this scale as a reference, the hardness of some common substances can be substituted for comparison: fingernail (2.5), copper penny (3.5), carpenter nail (4.5), knife blade (5.5), and steel file (6.5) When determining a mineral's hardness, one should be careful that a powdery residue is not mistaken for a scratch. Often cleaning the scratched area or

TABLE 9 THE MOHS' HARDNESS SCALE

1. Talc	6. Orthoclase
2. Gypsum	7. Quartz
3. Calcite	8. Topaz
4. Fluorite	9. Corundum
5. Apatite	10. Diamond

reversing the test by scratching one with the other will determine which substance is harder.

The specific gravity or density of a substance is a measure of its relative weight compared with that of an equal volume of water and is expressed in grams per cubic centimeter. This measurement is made by weighing the mineral first in air and then in water (Fig. 113). If the mineral weighs 2.0 units in air and 1.5 units in water, the difference of 0.5 unit is divided into the weight in air to yield a specific gravity of 4.0. Thus, if a mineral has a specific gravity of 4.0, it is four times as dense as water and has a density of 4.0. Therefore, density is numerically equal to specific gravity. Most nonmetallic minerals in the Earth's crust have densities between 2.5 and 3.0. Many common metallic minerals have densities over 5.0. The large difference between the densities of nonmetallic and metallic minerals becomes readily apparent when holding one of them in each hand.

Cleavage and fracture also indicate the structure of a mineral. A mineral has cleavage if it splits along a smooth plain parallel to a crystal face. Cleavage is described as poor, fair, good, perfect, or eminent. The various types of cleavage are determined by the number and direction of cleavage planes. Usually, a hammer blow or prying with a knife blade will determine whether a mineral has cleavage. Perhaps one of the most recognizable minerals with excellent cleavage is mica, which splits easily into thin, flexible sheets. Gemologists who cut precious gems and diamonds are well aware of cleavage patterns and study them very carefully before striking a large, valuable stone.

Minerals with little or no cleavage fracture along an irregular break when struck by a hammer. Most minerals fracture irregularly in one of several different ways, which aid in identification. A common type of fracture is conchoidal, or shelllike; it shows concentric rings, typically like those on broken glass. When metals break, they leave jagged edges and therefore have a hackly fracture. If a fracture is fairly smooth, it is an even fracture; if not smooth, it is an uneven or rough fracture. A fibrous or splintery fracture resembles broken wood. If a mineral breaks as clay does, it has an earthy fracture.

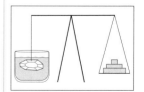

Figure 113 *The weight of a substance immersed in water is less than that in air; the comparison of the two weights constitutes the specific gravity.*

A flame test is used to detect the presence of certain chemical elements in a substance. A small amount of mineral is heated over a flame and its color is observed. Particular metals impart a characteristic color to the flame. For example, sodium produces a strong yellow color; copper produces a blue or green color, depending on the ore; chloride of copper produces a blue color; strontium and lithium produce a strong, intense crimson-red color; and potassium produces a violet color. A blowpipe is used to concentrate the heat of a flame so as to reduce a mineral to a residue, whose color is used for identifying the chemical elements present.

Other identifying features of minerals include taste, which can be tested with the tongue to identify halite; magnetism, which can be tested with a magnet to identify magnetite; electricity, often present when crystals are rubbed or heated, such as tourmaline; effervescence, which can be tested with a weak hydrochloric acid solution to identify calcite; fluorescence, which is the absorption and reradiation of ultraviolet light that can be tested with a black light to identify zinc and tungsten minerals; and radioactivity, which is the emission of particles by the decay of radioactive elements that can be tested with a radiation counter (Fig. 114) to identify uranium and thorium minerals.

Figure 114 *A scintillation counter is used to explore for radioactive minerals.*

(Photo courtesy URINCO)

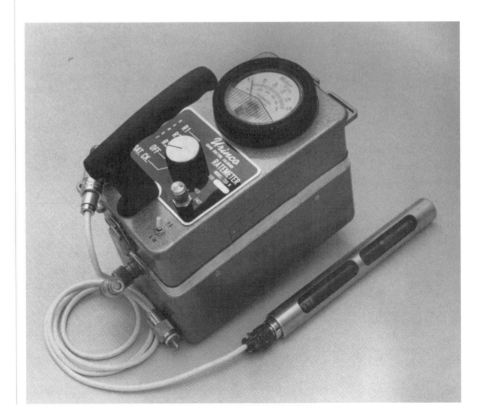

THE ROCK-FORMING MINERALS

Of the approximately 2,000 known minerals, only a few make up the majority of rocks found on the Earth's surface. Some occur in comparatively large quantities and are found over wide geographical areas, whereas others are relatively rare but are highly important economically. About 40 new minerals are identified each year, although most new finds are barely visible to the naked eye. Geologists classify minerals according to chemical composition and crystal structure. The main rock-forming minerals are silicates, of which iron, magnesium, sodium, calcium, or potassium has chemically combined with oxygen, silicon, and aluminum. These are the most common and widespread of the rock-forming minerals; Table 10 indicates the crystal abundance of rocks and minerals measured in percent volume of the entire crust.

Feldspars (Fig. 115) are the most abundant minerals and exist in two varieties: One is an aluminum silicate of sodium or calcium called plagioclase, and the other is an aluminum silicate of potassium called orthoclase and microcline. Rarer feldspars with barium and other metals are also known to occur. Feldspars are an important component of granite and sandstone. They also decompose into kaolinite, or ordinary clay. For this reason, feldspars are widely used in ceramics. Feldspars form monoclinic or triclinic crystals (refer to Fig. 111) with a white or pink to dark gray color, a pearly to vitreous luster,

TABLE 10   CRYSTAL ABUNDANCE OF ROCK TYPES AND MINERALS

Rock Type	Percent Volume	Minerals	Percent Volume
Sandstone	1.7	Quartz	12.0
Clays and shales	4.2	Potassium Feldspar	12.0
Carbonates	2.0	Plagioclase	39.0
Granites	10.4	Micas	5.0
Grandiorite, Quartz diorite	11.2	Amphiboles	5.0
Syenites	0.4	Pyroxenes	11.0
Basalts, Gabbros, Amphibolites, Granulites	42.5	Olivine	3.0
		Sheet Silicates	4.6
Ultramafics	0.2	Calcite	1.5
Gneisses	21.4	Dolomite	0.5
Schists	5.1	Magnetite	1.5
Marbles	0.9	Other	4.9

Figure 115 *Feldspar crystals in granite near Round Meadow, Tuolumne River, Yosemite National Park, Tuolumne County, California. The largest crystal is five inches in diameter.*

(Photo by G. K. Gilbert, courtesy USGS)

a hardness of 6, a density of 2.6 to 2.8, and two cleavages at nearly right angles. In addition, plagioclase has fine striations on a cleavage surface.

Quartz, which is the most common mineral after feldspar, forms six-sided prismatic crystals with a clear to gray color as well as various hues, such as pink for rose quartz. It is a prominent component of igneous, sedimentary, and metamorphic rocks. Some sandstones and quartzites are nearly 100 percent quartz. The St. Peter Sandstone mined for use in making glass is almost pure quartz. Large near perfectly formed quartz crystals found in pegmatites weigh up to a ton or more. Doubly terminated quartz crystals often found in limestone caves are sometimes cut as gemstones. Quartz crystals have a vitreous to greasy luster, a hardness of 7, a density of 2.6, and a conchoidal fracture. A type of quartz known as chalcedony is an important component of certain rocks, including flint, chert, and jasper.

Mica, which has perfect cleavage, has a clear variety called muscovite and a black variety called biotite. Micas are notable for being easily split into thin sheets or flakes. Large sheets of mica called "books" are used for heatproof windows and electrical insulation. The best and most perfect books of mica are mined in India. Muscovite, the most common mica, is mined commercially in many parts of the world. Micas are composed of oxides of alumina and silicon with other metals and often weather to produce clays. They form orthorhombic or hexagonal scalelike crystals with a vitreous to pearly luster, a hardness of 2 to 3, and a density of 2.7 to 3.0. Micas are common rock-forming constituents in igneous, metamorphic, and sedimentary rocks. Large six-sided mica crystals occur in pegmatites, weighing as much as 100 pounds.

Amphiboles are a large group of complex minerals composed of hydrated (combined with water) silicates of calcium, magnesium, iron, and aluminum. Hornblende is the most common amphibole and the main rock-forming mineral of the family. Hornblende decomposes by weathering to form clays. It forms long, slender six-sided prismatic orthorhombic or monoclinic crystals with a dark green to black color, a vitreous luster, a hardness of 5 to 6, a density of 2.9 to 3.2, and two cleavages at oblique angles. Amphiboles are common in basic or mafic (low silica and high iron or magnesium content) igneous and metamorphic rocks.

Pyroxenes are another large group of complex minerals closely related to amphiboles composed of silicates of calcium, magnesium, iron, and aluminum. They are often found as primary minerals in igneous rocks. Augite is the most common pyroxene and the main rock-forming mineral of the family. It forms stubby eight-sided prismatic monoclinic or orthorhombic crystals (refer to Fig. 111) with a dark green to black color, a vitreous to dull luster, a hardness of 5 to 6, a density of 3.2 to 3.6, and two cleavages at right angles. Pyroxenes are common in nearly all basic (of low silica content) igneous and metamorphic rocks and are sometimes found in meteorites.

Zeolites, although not major rock-forming minerals, are widely distributed. All are chemically related to the feldspars and are composed of hydrous silicates of aluminum with other metals. The water is held so loosely that zeolites tend to boil and bubble when heated; that is why they are called "boiling stones." They are often found in lava flows with cavities or vesicles filled with zeolite crystals that formed when water boiled away as the basalt cooled. Most zeolites have a framework of small tetrahedral molecular units comprising triangular pyramids in which four oxygen atoms enclose a central aluminum or silicon atom. The most common variety of mineral in zeolites is stilbite, which has monoclinic crystals with a white, yellow, brown, or red color; a glassy or pearly luster; a hardness of 3.5 to 4; and a density of 2.1. Because zeolites readily exchange ions of calcium and sodium, they are used in water softeners. They are also used in oil refineries to break down large oil molecules into smaller, more useful products.

Garnets, which are better known as gems than as rock-forming minerals commonly constitute a small but conspicuous ingredient of igneous and metamorphic rocks. They form a group of closely related minerals, mostly silicates of various metals, including calcium, magnesium, iron, and aluminum. All varieties form isometric crystals up to four inches wide with 12 or 24 sides or a combination of the two. Colors range from deep red for the gemstone pyrope to black, brown, yellow, or brilliant green for the gem demantoid, which resembles an emerald in color. The mineral has a glassy luster, a hardness of 7, and a density of 3.4 to 4.3. Only a small percentage of garnets are of gem quality and most are used commercially as abrasives.

Olivine is the most common member of one group of silicates. It is an iron-magnesium silicate and the simplest of the dark minerals. Olivine is found in igneous rocks rich in magnesium and low in quartz such as basalt and gabbro. It has small, sugary grains, and crystals are relatively rare, although occasionally some up to several inches long have been found. Clear yellow-green varieties are cut as the gemstone peridote. Olivine has a distinctive olive green color (hence its name), a vitreous luster, a hardness of 6.5 to 7, a density of 3.2 to 3.5, and a conchoidal fracture. Olivine is an important constituent of many basic or mafic igneous and metamorphic rocks.

Chlorite is a hydrous iron-magnesium aluminum silicate with monoclinic crystals. It often results from the alteration of rocks rich in pyroxenes, amphiboles, and biotite. Chlorite is usually colored some shade of green but can vary from white to brown and black. It has perfect cleavage and can split into thin sheets similar to micas. The sheets are flexible similar to mica but unlike mica they are inelastic and unable to spring back. Chlorite has a glassy to pearly luster, a hardness of 2 to 2.5, and a density of 2.6 to 2.8. It is usually found in metamorphic rocks, such as low-grade schists and greenstones, but also occurs frequently in cavities of basic igneous rocks.

Serpentine is chemically similar to the mineral chlorite, with a mixture of magnesium and iron-aluminum silicates and water. In addition, it might contain small amounts of nickel. One variety known as chrysolite is a fibrous form of serpentine called asbestos that is used for insulation. Serpentine is translucent to opaque, varies in color from cream white through all shades of green to black, and has a greasy or waxlike luster, a hardness of 2.5 to 4, and a density of 2.6. Serpentine, so named because of its mottled green color, is soft and easily polished and can be worked into a variety of ornaments and decorative objects. An especially beautiful deep green form used for indoor decoration is called verde antique marble, although it is not a true marble, which is actually metamorphosed limestone.

Calcite is a calcium carbonate and the most common mineral in limestone, where often great masses of calcite crystals occur. It is the most widespread of the carbonate minerals and one of the most interesting minerals because of its many and varied crystal forms. It has dogtooth or flat six-sided crystals with a clear to white color, but impurities can produce the colors: yellow, green, orange, or brown. Calcite grades into dolomite with the substitution of magnesium for calcium, which makes it significantly harder. Calcite crystals have a vitreous or dull luster, a hardness of 3, a density of 2.7, and excellent cleavage in three directions. Another form of calcite, called travertine, is common in hot carbonate springs (Fig. 116) and in caves. Calcite also forms in igneous vein deposits called carbonatites. It effervesces strongly in dilute hydrochloric acid.

Figure 116 *Hot carbonated spring water undercuts bedded travertine deposits at Yellowstone National Park, Wyoming.*

(Photo by K. E. Barger, courtesy USGS)

Gypsum is a hydrated calcium sulfate that forms by the evaporation of seawater under arid conditions. It has tabular or fibrous monoclinic crystals with a clear or white color, a vitreous to pearly luster, a hardness of 2, a density of 2.3, and perfect cleavage with flexible but nonelastic flakes. It is mined extensively from thick beds for building materials such as plaster and wallboard. Plaster artifacts made from gypsum include containers, sculptures, and ornamental beads that were discovered in the Near East, dating as far back as 14,000 years ago, long before the discovery of pottery. Single gypsum crystals are found in black shales. A compact, massive form known as alabaster is used for carving ornamental objects.

Halite is a sodium chloride commonly called rock salt. It is colorless when pure but usually occurs in discolored shades of yellow, red, gray, or brown. It is transparent to translucent and brittle, with excellent cleavage parallel to crystal faces. It occurs in granular, fibrous, or crystalline masses recognized by cubic crystals, a hardness of 2.0 to 2.5, and a density of 2.3. Halite is extracted from formerly shallow, stagnant pools of seawater called brines that have evaporated, forming bedded deposits. It follows gypsum and anhydrite in the sequence of precipitation of salts from seawater. Salt has been mined extensively throughout the world since ancient times and remains a valuable commodity.

Common ore-forming minerals include hematite, which is distinguishable by its blood-red color and is the most abundant ore of iron. Pyrite, called "fool's gold" because of its golden appearance, is known for its cubic, brassy crystals. Chalcopyrite, another "fool's gold," is the most abundant ore of copper and often an ore of gold or silver. Argentite, the most important ore of silver, is lead-gray in color and usually occurs in solid masses or coatings on other minerals. Galena (Fig. 117) with its heavy, gray cubic crystals is the most important ore of lead. Sphalerite forms yellowish to brownish cubic crystals with six perfect cleavages and is the most important ore of zinc.

Cassiterite forms brown or black tetragonal pyramidlike crystals and is practically the sole ore of tin. Bauxite, which occurs in rounded grains or earthy masses with a range of colors, including white, gray, yellow, or red, is the most important ore of aluminum. Cinnabar, easily identified by its vermilion-red color, occurs in finely granular masses and is the only important ore of mercury. Ilmenite, found as thick, flat crystals; thin plates; or compact masses colored iron black, often forms the "black sands" of many beaches and is the major ore of titanium. Uraninite, also called pitchblende because of its resemblance to pine pitch, is the chief source of uranium.

Pseudomorphs, meaning "false forms," are some of the most fascinating members of our planet's mineral kingdom. A crystal can have the shape of one mineral but actually possess the chemical makeup of an entirely different type. With increased heat or pressure or infiltration by water or acid, minerals might

Figure 117 *A partly oxidized ore from the Lee mine, showing galena (gn), Darwin quadrangle, Inyo County, California. The constituents include bindhemite (bn), cerrusite (ce), chalcedony (cl), calcite (ct), galena (gn), and heminophite (hm).*

(Photo by W. E. Hall, courtesy USGS)

adapt by changing into different minerals more suited to the environment and therefore more stable under the new conditions.

The most common pseudomorphs are those that change their chemical composition but retain the shape of the original mineral. For example, cuprite composed of copper sulfide can lose its sulfur and become pure copper, with the new mineral retaining the shape of the original. If water invades the environment of bright blue asurite, it transforms into the brilliant green mineral malachite. Even petrified wood is considered a pseudomorph because water infiltrates a tree's cells and replaces the organic matter with various types of quartz such as jasper and agate, while maintaining the original structure of the tree.

ORE DEPOSITS

Ores are naturally occurring materials from which valuable minerals are extracted. We have been blessed with a world rich in mineral wealth and have barely scratched the surface in our quest for ore deposits. Improved techniques in geophysical and geochemical processes and mineral exploration have helped keep resource supplies up with rising demand. As improved exploration techniques become available, future supplies of minerals will be found in yet unexplored regions. Immense mineral resources lie at great depths, awaiting the mining technology to transport them to the surface for use in industry.

Mineral ore deposits form very slowly, taking millions of years to create an ore sufficiently rich to be suitable for mining. Certain minerals precipitate over a wide range of temperatures and pressures. They commonly occur together with one or two minerals predominating in significantly high concentrations to make their mining profitable. Extensive mountain building activity, volcanism, and granitic intrusions have provided vein deposits of metallic ores.

The dominance of hydrothermal (hot water emplaced) deposits as a major source of industrial minerals has stimulated intense study of their genesis for over a century. Toward the turn of the 20th century, geologists found that hot springs at Sulfur Bank, California, and Steamboat Springs, Nevada (Fig. 118), deposited the same metal-sulfide compounds that are found in ore veins. Therefore, if the hot springs were depositing ore minerals at the surface, hot water must be filling fractures in the rock with ore as it moves toward the surface. The American mining geologist Waldemar Lindgren discovered rocks with the texture and mineralogical characteristics of typical ore veins by excavating the ground a few hundred yards from Steamboat Springs. He proved that many ore veins formed by circulating hot water called hydrothermal fluids. The mineral fillings precipitated directly from hot waters percolating along underground fractures.

Hydrothermal ores originate when a gigantic subterranean still is supplied with heat and volatiles from a magma chamber. As the magma cools, silicate minerals such as quartz crystallize first, leaving behind a concentration of other elements in a residual melt. Further cooling of the magma causes the rocks to shrink and crack, allowing the residual magmatic fluids to escape toward the surface and invade the surrounding rocks to form veins. The rocks surrounding a magma chamber might be another source of the minerals found

Figure 118 Steam fumaroles at Steamboat Springs, Nevada.

(Photo courtesy USGS)

in hydrothermal veins, with the volcanic rocks acting only as a heat source that pumps groundwater into a giant circulating system. Cold, heavier water moves down and into the volcanic rocks carrying trace amounts of valuable elements leached from the surrounding rocks. When heated by the magma body, the water rises into the fractured rocks above, where it cools, loses pressure, and precipitates its mineral content into veins.

Another type of mineral ore emplacement called massive sulfide deposits originated on the ocean floor at midocean spreading centers and occur as disseminated inclusions or veins in ophiolite complexes that were exposed on dry land during continental collisions. Among the most noted deposits are the 100-million-year-old Apennine ophiolites, which were first mined by the ancient Romans. Massive sulfide deposits are mined extensively in various parts of the world for their rich ores of copper, lead, zinc, chromium, nickel, and platinum.

Two metals on opposite extremes of the hydrothermal spectrum are mercury and tungsten. All belts of productive deposits of mercury are associated with volcanic systems. Mercury is the only metal that is liquid at room temperature. It forms a gas at low temperatures and pressures, and therefore much of the Earth's mercury is lost at the surface through volcanic steam vents and hot springs. Tungsten, by comparison, is one of the hardest metals; that property makes it valuable for hardening steel. It precipitates at very high *temperatures* and *pressures,* often at the contact between a chilling magma body and the rocks it invades.

Economic deposits of iron ore are found on all continents. Layered deposits of iron oxide cover huge regions, such as the Lake Superior region of North America and the Hamersley Range of Western Australia. The Mesabi Range of northeast Minnesota is the major supplier of iron ore for the United States. The ore occurs in a banded iron formation laid down over 2 billion years ago. The Clinton iron formation is the chief ore producer for the Appalachian region. The iron occurs in an oolitic ironstone deposited over 400 million years ago. The El Laco mine on the border between Chile and Argentina is a rarity among iron mines. The ore body is a large lava flow consisting almost entirely of the iron minerals hematite and magnetite.

Important lead and zinc hydrothermal deposits exist in the tri-state region (Missouri, Arkansas, and Tennessee) of the Mississippi River valley near Missouri. Copper, tin, lead, and zinc ores concentrate directly by magmatic activity, forming hydrothermal vein deposits. Rich deposits of copper, lead, zinc, silver, and gold exist in the Cordilleran mountain regions of North and South America. Zambia's great copper belt is estimated to contain a quarter of the world's copper. A huge copper belt 100 miles long and three miles wide and some 2 billion years old exists on the Keweenaw peninsula in the Lake Superior region. A variety of other metallic deposits lie in the mountains of southern Europe and in the mountain ranges of southern Asia as well. The largest nickel deposit, located at Sudbury, Canada, is truly out of this world, for it is thought to have originated from a massive meteorite impact about 1.8 billion years ago.

Gold is mined on every continent except Antarctica. In Chile, gold and silver were mined from the eroded stumps of ancient volcanoes. In Africa, the best gold deposits are in rocks as old as 3.4 billion years. In North America, the most productive gold mines are in the Great Slave region of northwestern Canada, where over 1,000 deposits are known. These gold deposits are found in greenstone belts that were invaded by hot magmatic solutions from the intrusion of granitic bodies; the gold occurs in veins associated with quartz.

Chromium has only one ore, namely, chromite, although it occurs in about a dozen minerals. Half the world's production of chromium is in South Africa, which is also responsible for much of the global diamond production. The diamonds are disseminated in a volcanic structure called a kimberlite pipe that resembles a funnel reaching deep into the Earth's mantle (see Figure 131). Most diamond-bearing kimberlite pipes of South Africa are about 100 million years old, although the diamonds they contain formed billions of years ago. The major platinum deposits of the world include the Bushveld Complex of South Africa and the Stillwater Complex of Montana.

Sulfur is one of the most important nonmetallic minerals. It occurs in abundance in sedimentary and evaporite deposits, with volcanoes contributing only a small proportion of the world's economic requirements. The largest volcanic sulfur mines are in northern Chile. The open pit mine atop Aucanquilcha Volcano has the distinction of being the highest mine in the world, at some 20,000 feet in elevation. The mine is situated in the heart of a complex andesite volcano, and the entire central part contains a rich sulfur ore.

Valuable reserves of phosphate used for fertilizers are mined in Idaho and adjacent states. Evaporate deposits in the interiors of continents, such as the potassium deposits near Carlsbad, New Mexico, indicate these areas were once inundated by ancient seas. Thick beds of gypsum used in the manufacture of plaster of paris and dry wall board also were deposited in the continental interiors. Nonmetallic minerals such as sand and gravel, clay, salt, limestone, gypsum, and phosphates are mined in great quantities throughout the world.

On the ocean floor, the most promising mineral deposits are manganese nodules (Fig. 119). Most manganese nodules are particularly well developed in

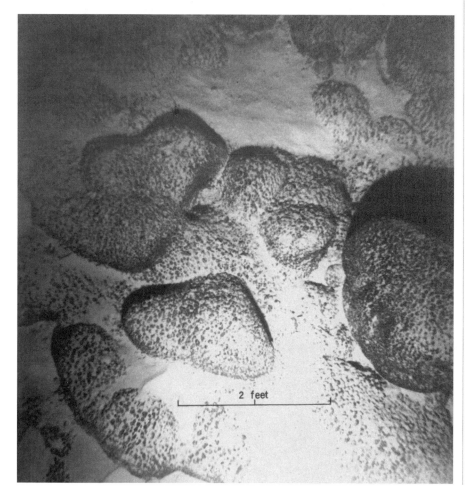

Figure 119 *Manganese nodules on Sylvania Guyot, Marshall Islands, at a depth of 4,300 feet.*

(Photo by K. O. Emery, courtesy USGS)

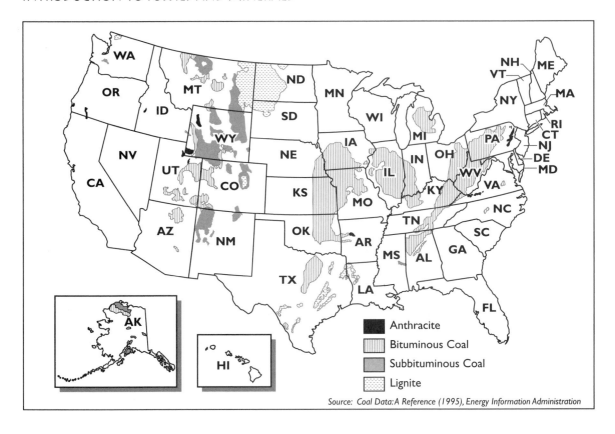

Source: Coal Data: A Reference (1995), Energy Information Administration

Figure 120 Coal deposits in the United States.

deep, quiet waters far from continental margins and active volcanic zones. Concentric mineralized layers accumulate over millions of years until the nodules reach about the size of a potato, giving the ocean floor a cobblestone appearance. A ton of manganese nodules contains about 600 pounds of manganese, 29 pounds of nickel, 26 pounds of copper, and about seven pounds of cobalt. But their location at depths approaching four miles makes extraction on a large scale extremely difficult.

Abundant coal reserves are found in the United States (Figs. 120, 121), Canada, South Africa, and Asia. Large quantities of oil and gas are found in the Middle East, the Gulf Coast region, the Rockies, the North Slope of Alaska (Fig. 122), and the North Sea. Huge untapped reserves of oil are in oil shale deposits in the western United States (Fig. 123) with a potential global oil content exceeding that of all other petroleum resources in the world.

The seemingly insatiable appetite for fossil fuels and ores to maintain a high standard of living in the industrialized world as well as to improve the standard of living in developing nations could very well lead to the depletion of known petroleum and high-grade ore reserves by the middle of this cen-

Figure 121 *The Big Elk coal bed, King County, Washington.*

(Photo by J. D. Vine, courtesy USGS)

Figure 122 *Oil well drilling on Alaska's North Slope, Barrow District, Alaska.*

(Photo by J. C. Reed, courtesy USGS)

tury. Then, low-grade deposits would have to be worked, potentially dramatically increasing the cost of goods and commodities. Only by conserving natural resources will the wealth of the Earth be preserved for future generations.

The most valued minerals, gemstones and precious metals, will be discussed in the next chapter.

8

GEMS AND PRECIOUS METALS

MINERALS OF GREAT VALUE

Gems are highly prized minerals that have a common appeal to all cultures, and their legacy extends to prehistoric times. As far back as 20,000 years ago, our ancestors, the Cro-Magnon, ornamented their bodies with lavish strings of beads made from ivory, seashells, and gemstones. As is the custom today, the use of jewelry was determined by fashion and reflected one's rank in society. Gems were also thought to have mystical powers. Primitive cultures and even some modern people believe that gems and crystals have the ability to heal.

All gems are purer, clearer, or more crystalline forms of common minerals, which are generally less beautiful and less spectacular specimens. Diamonds, emeralds, rubies, and sapphires stand out as true gems, whereas other stones are classified as semiprecious and ornamental stones. Scarcity and fashion usually determine the value of a gem, but other factors include luster, transparency, color, and hardness. Luster depends on the way light is reflected by the mineral. The transparent gems also refract or bend light and are cut specifically to turn the light back to the eye of the observer. Color, which is essential in some gems and incidental in others, can add or detract greatly from a gem's value. Hardness is also important, and the harder the gem, the better it resists scratching of its highly polished surfaces.

Growing in popularity along with the gemstones were the precious metals gold and silver, and much blood has been spilt down through the ages for the attainment of these commodities. The plundering of Inca gold and silver by the 16th-century Spaniards was largely responsible for the downfall of the greatest empire in South America. As poetic justice of a sort, the inflation of large amounts of Inca gold and silver into its economy eventually brought down the Spanish empire as well. The forty-niners' mad scramble for gold in California quickly swelled its population, gave the state a unique character, and allowed it to enter the Union much sooner than expected. The miners worked placer deposits, known as "poor man's deposits," where, besides gold, gemstones were found.

QUARTZ GEMS

The quartz gems are the best known among the semiprecious stones and produce a greater variety of gemstones than any other mineral. The transparent varieties possess a rainbow of colors from clear to yellow, blue, violet, green, pink, brown, and black. Rose quartz, named for the color of the flower, has a subtle shade of pink due to the presence of manganese. Smoky quartz derives its brown color from the presence of small amounts of radioactive elements such as radium, which irradiate silicon atoms. Quartz cat's-eyes are formed when quartz crystallizes around preexisting minerals, sometimes altering them into unusual specimens. The translucent or opaque quartz gems, which are grouped under the name *chalcedony,* have a wide array of colors and forms, and some have a banded, striped, or mottled appearance. Doubly terminated quartz crystals often found in limestone caves are sometimes cut as gemstones.

Amethyst is the most valuable of the quartz gems, and until the 18th century, it ranked among the most precious of stones. It lost much of its value after the great Brazilian deposits were found. Amethyst contains traces of iron or manganese finely dispersed throughout the crystal, giving it a spread of colors from orchid pink to regal purple, hues not seen elsewhere in the entire realm of gems. The deeper-colored specimens are cut as gems that are highly prized. The stone was once believed to protect its wearer from drunkenness. (Of course, he had to abstain from drinking alcohol to gain that protection.)

Opal is perhaps the most famous of the quartz gems and is renowned for its vivid flashes of colored light, typically red, orange, yellow, and green, colors no other gemstone possesses. Opals receive their fiery colors from crystallized silica colloids (microscopic grains) that happen to be the right size for scattering light. The silica gem occurs in small veins, or irregular masses of rounded forms, and contains varying amounts of water, typically 3 to 9 percent by volume. It is often cut in the rounded shape called cabochon. Common opal is milky white, yellow-green, to brick red and is somewhat translucent, glassy, or resinous. Fire opal is so named because of its flamelike colors. The play of colors in opal is produced by the diffraction of light rather than by absorption. The flashes can spread evenly across the gem's surface or become irregular points of light as the gem is rotated. The scintillations probably arise from the manner in which opal was deposited, involving layers of microscopic spherules of hydrated silica that accumulated in thermal springs. This form of silica also comprises the skeletons of diatoms (Fig. 124) and siliceous sponges. Black opal from Nevada is found replacing other minerals in fossil wood, seashells, and dinosaur bones.

Agate is known for its bands of alternating layers formed by intermittent deposits of silica from solution into irregular cavities in volcanic rocks. The concentric wavy patterns are derived from irregularities in the walls of the cavity. A number of varieties of agate are characterized by peculiarities in the shape and color of the bands. Moss agates, found along the Yellowstone River and in adjacent regions, show magnificent landscapes (Fig. 125). The fernlike crusts or dendrites are composed of pyrolusite, an ore of manganese that forms as inclusions in moss agates. Petrified wood is usually agatized, as silica in groundwater solutions replaces woody tissues. Many legends abound concerning the healing power of agate, and it was once used to banish fear and protect against epilepsy.

Onyx is a variety of chalcedony similar to banded agate. It has even, parallel bands usually colored black and white, brown and white, or red and white. It is largely used for making cameos (small sculptures carved on stones) because the design and background can be cut so as to appear in different-colored layers. The best cameos were produced by the ancients, and a revival of the art took place in the mid-19th century, when rich deposits of onyx were discovered in South America. Onyx marble is a compact, usually translucent variety of calcite and sometimes aragonite that is similar in appearance to true onyx. If composed of parallel-banded travertine, it makes an excellent decorative or architectural material. It is usually deposited from cold water solutions, often found in the form of stalactites and stalagmites in caves (Fig. 126).

Jasper is a cryptocrystalline quartz gem composed of microscopic crystals. It is a chalcedony, a form of quartz closely related to agate, and is composed of a large variety of chemical elements, which in turn give it various colors and designs. The colors range from red to yellow, brown, or any combination of the three. The red varieties usually contain hematite, an ore of iron. Sometimes it is found banded with several different-colored stripes. Over time, jasper can further grade into chert. The ancients attributed

Figure 124 *Miocene age diatoms from the Choptank Formation, Calvert County, Maryland.*

(Photo by G. W. Andrews, courtesy USGS)

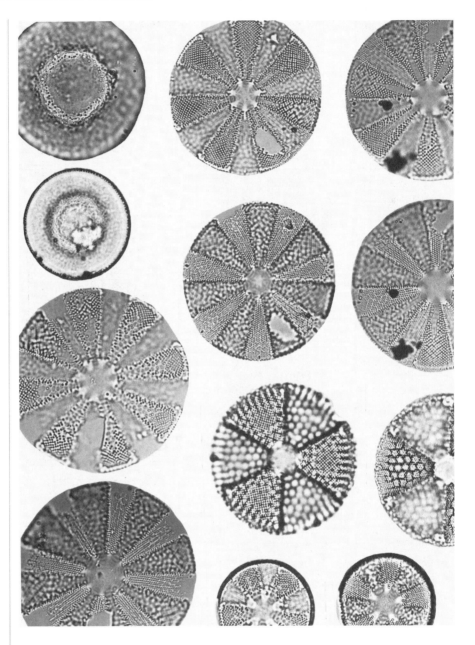

many medicinal properties to this stone, and even as late as the 17th century people still believed that jasper hung around the neck could cure stomach disorders.

Bloodstone, also called heliotrope (from the Greek *helio,* meaning "sun"), was named the "gem that turns the sun"; quantities of this stone were found near the ancient Egyptian city of Heliopolis. It is a dark green quartz with small spots of red jasper that resemble drops of blood. The gem was greatly prized in the Middle Ages and used for making sculptures representing violent death or martyrdom. Bloodstone was also believed to be capable of stopping hemorrhages and causing tempests.

TRANSPARENT GEMS

The transparent gems are known for their striking luster and brilliance along with their color and hardness. Most transparent gems are oxides of aluminum, beryllium, or magnesium, with a few containing silica. Some transparent gems such as ruby and sapphire are identical in mineral composition and only differ in color. The value of these stones is determined largely by their transparency, lack of flaws, brilliancy of color, and size. The corundum gems, including ruby and sapphire, are all rare forms of alumina (Al_2O_3) that vary widely in color. Green, purple, and yellow corundum are known as oriental emerald, oriental amethyst, and oriental topaz. In addition, synthetic rubies, sapphires, emeralds, and similar gems of large size and

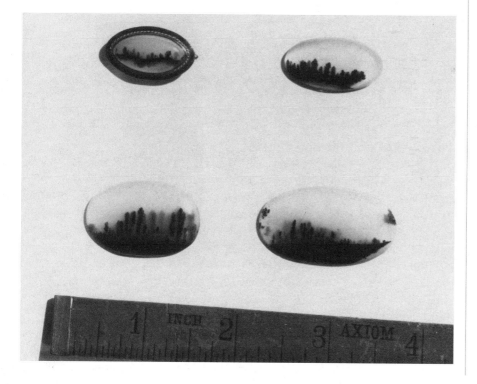

Figure 125 *Agates covered by moss-formed landscapes are found along the Yellowstone River and in adjacent regions of Montana.*

(Photo by D. B. Sterrett, courtesy USGS)

Figure 126 *Hattin's Dome and Temple of the Sun at Carlsbad Caverns National Park, Eddy County, New Mexico.*

(Photo by W. T. Lee, courtesy USGS)

fine quality are made by fusing fine alumina with appropriate mineral pigments under high temperatures.

Ruby is the most valued of all gemstones, and the deep red varieties are prized even more than diamonds. It is a vivid red gem of the mineral corundum, which is second in hardness to diamond on the Mohs' hardness scale. The color, attributed to the presence of a small amount of chromium, varies from deep red to pale rose red, and in some stones it shows a tinge of violet. The oriental ruby, mined mostly in Burma, has very limited distribution, which accounts for its extraordinary high value. The ruby was considered by the Hindus the king of the precious stones. It was also thought to bestow good fortune when worn on the left side of the body.

Sapphire is the serene blue variety of corundum and is essentially the same mineral as ruby, differing only in color and being slightly harder. It also occurs in vivid green, violet, and yellow hues. The color of the ever-popular rich blue sapphire is attributed to the presence of oxides of chromium, iron, or titanium. The Montana variety has a peculiar electric blue color. Star sapphires are unusual in that they reflect light in a figure of a six-pointed star.

Emerald, known for its deep green color, is a gem of the mineral *beryl* and should not be confused with the oriental emerald, which is actually an emerald-colored sapphire. Beryl is an important ore of beryllium as well as a major gemstone. The crystals are hexagonal and are principally found in granite pegmatites. When colored green, beryl is emerald, when colored blue or blue-green it is aquamarine, and when colored pink it is morganite. Emerald colors vary from light green to dark green and are due to the presence of chromium. Compared to the other transparent gems, the emerald is relatively soft, only slightly harder than quartz. Emeralds were mined in Egypt as early as 1650 B.C., and Cleopatra's mines, located on the Red Sea coast east of Aswan, yielded precious emerald gems to adorn the queen of Egypt. Emeralds were once thought to have therapeutic value, curing such afflictions as poisoning, diseases of the eye, and the possession of demons.

Zircon (Fig. 127) is one of the most extraordinary gems. It is fairly common in igneous rocks, but rare as a gemstone. Zircon crystals found in granite are enormously resistant to erosion and tell of the earliest history of the Earth, when the crust first formed some 4.2 billion years ago. Zircon is also

Figure 127 *Zircons from rare-earth zone, Jasper Cuts area, Gilpin County, Colorado.*

(Photo by E. J. Young, courtesy USGS)

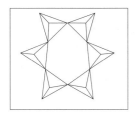

Figure 128 *The structure of a tourmaline crystal.*

an important ore for zirconium, hafnium, and thorium. A rich golden brown variety is the most magnificent of all gemstones possessing this color. The natural colors range from clear varieties, which are favorite substitutes for diamonds, through shades of yellow and brown to deep brownish red. Brown zircon crystals can be altered to a rich blue variety by heating them in the absence of air, a potential that makes them more highly prized gems.

Tourmaline, often called the "rainbow gem," displays the widest range of exquisite colors of all the gems. The crystals are often long and are unique in having curved triangular cross sections. Not only does tourmaline appear in almost every color of the rainbow, but a single crystal might be half one color and half another or show three different colors resembling a candy cane. The profusion of colors is due to a complex chemical composition, including perhaps a greater variety of chemical elements than any other mineral. The most valuable colors are a clear ruby red and a bright sapphire blue. Tourmaline is very common in pegmatites, where it sometimes occurs in crystals of enormous size. Because of its unique crystal structure (Fig. 128), tourmaline has a strange electrical property. It becomes positively charged at one end and negatively charged at the other end of the crystal when heated. The static electricity attracts such objects as small bits of paper similarly to the way a comb drawn through the hair does.

Garnet is a rare gem of silicates of various metals with a typically red, brown, yellow, white, green, or black color and a glassy to resinous luster. Garnets are a common group of closely related silicate minerals, containing calcium, magnesium, iron, and aluminum, combined with silicon and oxygen. Crystals are usually 12- or 24-sided or any combination to yield 36 and 48 sides. Demantoid is a brilliant green variety that resembles an emerald in color. Pyrope, sometimes called precious garnet, is deep red and used in jewelry. Perfect specimens are found in South African "blue earth" in kimberlite pipes associated with diamonds and make exceedingly fine gems.

Staurolite is an iron-aluminum silicate often found with garnets in metamorphic rocks such as schists, phyllites, and gneisses and in pegmatites. An unusual inclusion of staurolite mineral in a South African diamond implies that it came from a piece of continental crust that was recycled through the mantle. Staurolite is brown to black and almost always occurs as stubby prismatic crystals usually an inch or less long, but in excellent specimens they are up to two inches long. Twinning is common, and when occurring at right angles the twin crystals form a perfect cross (Fig. 129). These crystals, called fairy crosses, are broken or weathered from the bedrock and are sold as charms or souvenirs. Transparent crystals are rare and are cut as gemstones.

Peridot is a clear yellow-green gem of the mineral olivine, which is the most common member of a group of silicates found in many igneous rocks. It is an iron-magnesium silicate and the simplest of the dark minerals. Olivine is found in igneous rocks rich in magnesium and low in quartz such as basalt and gabbro. It has small, sugary grains, and crystals are relatively rare, although occasionally some up to several inches long have been found. *Peridotite* is the rock name for this mineral, which is significant because a very similar rock type is mined from volcanic structures called kimberlites for diamonds, indicating the mineral originated within the Earth's mantle.

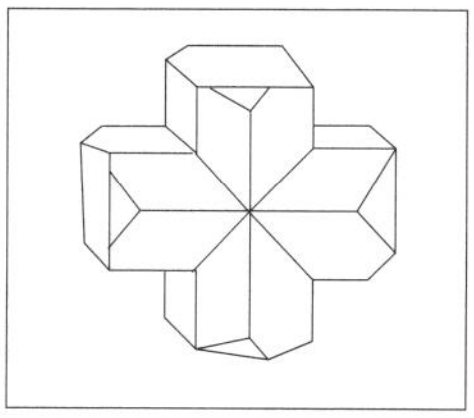

Figure 129 Twinning of a staurolite crystal showing a perfect cross.

Topaz is commonly a yellow gemstone but ranges in color from pale yellow to brown. It has been very popular for jewelry since the 16th century. The rare pink topaz is much admired, especially the deeper hues. Topaz has a remarkably slick surface and a slippery feeling that distinguishes it from other minerals. Some large crystals, weighing upward of 600 pounds, have been found in pegmatites, with colors varying from yellow to blue, green, violet, and colorless. The yellow, yellow-brown, and blue-green varieties often occur in beautiful crystals that make valuable gems. The most important commercial source of the gemstone is Brazil. The mystical powers of topaz are believed to increase when the moon is new or full in the sign of Scorpio, at which time the wearer is able to receive strong impressions from astral sources.

OPAQUE GEMS

The opaque gems include representatives of metallic ores and rock-forming minerals, and with the exception of jade they are used mainly as ornamental stones. Some opaque gems such as obsidian, a volcanic glass, and jet, an extremely hard form of coal, are better classified as rocks. (An unusual mineral that rightfully belongs to this group is the pearl because, as with all gems, it is highly prized for its beauty.)

Turquoise is a sky blue gemstone that has been used as an ornament since the dawn of civilization. Jewelry made from turquoise has been found in Egyptian and Sumerian tombs, dating back to the fourth millennium B.C. Because of its softness (slightly less than 6 on the Mohs' hardness scale), turquoise was easily worked with the primitive tools available in ancient times. The demand for the gem in the United States has been high because of the popularity of turquoise jewelry made by Native Americans in the West. For ages, the gem has been admired by the Navajos, who mined it long before the arrival of the Europeans. Most deposits were in New Mexico, Arizona, Colorado, and Nevada. The mineral is associated with copper and occurs in nuggets or in veins. Sometimes tiny veins of clay and iron oxide crisscross the stone, giving it a much greater appeal.

Jade has been highly regarded since ancient times for its pleasant green color and versatility. It is a waxy or pearly mineral that is usually green but is also yellow, white, or pink. Unlike other gems, which are usually varieties of a single mineral, jade has complex mineralogical attributes. It occurs in two varieties that are of different chemical composition but are similar in appearance. Jadeite is a pyroxene, and the light, translucent emerald green form is considered a precious stone. It is regarded as the more valuable of the two jades because it has a richer appearance and possesses a greater variety of colors. Nephrite is an amphibole and the more common of the two jades. Both varieties of jade have been carved into ornaments and implements since antiquity.

Moonstone, which is a variety of plagioclase called albite, is valued as a gemstone because of its bluish white or pearly opalescence. It receives its name from a moonlike silvery white sheen that changes on the surface as the light changes. Almost all moonstone of commercial value is from mines in Sri Lanka (Ceylon) off the southern tip of India. One belief is that knowledge of the future can be obtained by holding a moonstone in the mouth under a waning moon.

Malachite is a common ore of copper, and because of its conspicuous bright green color, it is a useful guide in copper prospecting. It is often found together with azurite, which forms deep blue crystals. Both occur in smooth or irregular masses in the upper levels of mines. Malachite sometimes possesses crystals with a glassy luster but usually occurs in fibrous rounded masses with a silky luster. The compact, deep-colored stones make beautiful ornaments when cut and polished. Malachite is also fashioned into urns, bowls, and a great variety of art objects. In the Middle Ages, malachite was especially treasured as a protection against the "evil eye."

Lapis lazuli (lazurite) has been in high favor as an ornamental stone for thousands of years. Deposits in a very remote part of Afghanistan were mined more than 6,000 years ago. It has an intense purplish blue color, which, unlike that of many gems, does not fade in sunlight. The stone is used to make a variety of art objects and can be cut into a gemstone, although its softness does not make it entirely suitable for this purpose. An old tradition holds that the Ten Commandants were inscribed on two slabs of this stone.

Serpentine is another ornamental stone, composed of a mixture of magnesium and iron-aluminum silicates and water. It also might contain small amounts of iron or nickel. Serpentine is translucent to opaque with a greasy or waxy luster and varies in color from cream white through all shades of green to black. Serpentine, so named because of its mottled green color, like that of a serpent, is soft and easily polished and can be worked into a variety of ornaments and decorative objects. An especially beautiful deep green form used for indoor decoration is called verde antique marble.

Pearl, known as the queen of gems, has been treasured by civilizations in all parts of the world for thousands of years. Even to this day, natural pearls are among the most prized of gemstones. Pearls are formed by a number of marine and freshwater mollusks such as clams and oysters when sand or some other particle irritates the animal's mantle that lines the inside of the shell (Fig. 130). As a result, layers of aragonite grow year by year into a spherical concretion with a captivating iridescent luster. Pearls are also grown commercially by artificially implanting seeds inside oysters. Pearls might also be gold, pink, red, or black. The pearl was once esteemed as the emblem of purity, innocence, and peace.

Amber is a metamorphosed resin from ancient tree sap buried deeply under high pressures for up to 100 million years or more. The most extensive deposits occur along the Baltic Sea and represent extinct flora of Tertiary age. Most amber occurs in shades from yellow to brown, but some specimens are opaque white. Often insects and even small tree frogs become trapped and preserved in amber. The process by which resin preserves tissue so that it retains its original size is still a mystery. Air bubbles captured in amber from the Cretaceous period indicate a much higher atmospheric oxygen content, which might have contributed to the giantism of the dinosaurs. Even deoxyribonucleic acid (DNA), which constitutes the molecules of heredity, has been extracted from extinct species trapped in amber. Wonderfully preserved specimens in amber make it a valuable gemlike stone.

DIAMONDS

The diamond is the most important of the gemstones formed when pure carbon is subjected to high pressures and temperatures of the Earth's deep inte-

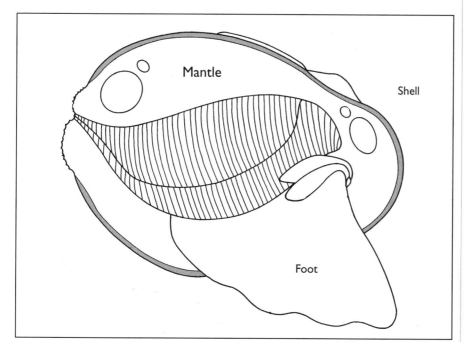

Figure 130 A cross section of a clam showing mantle area, where a pearl often forms.

rior. Microdiamonds have also been produced by meteorite impacts, whereby the high-pressure collisions fused carbon into diamond dust. Diamonds with small black inclusions of foreign minerals are used as evidence that continental material can get trapped in subduction zones and pulled deep down into the mantle, where high pressures compress carbon atoms into tight crystals. Diamonds also can be produced synthetically in the laboratory by compressing carbon to extreme temperatures and pressures.

Diamond crystals are usually six-sided or eight-sided, although they are often found in distorted and irregular shapes. A diamond's value depends on its hardness and its brilliance, derived from a high index of refraction, or bending of light rays, which accounts for its so-called fire produced by cutting the stone in such a way that light is refracted and highly dispersed as it passes through the gemstone. The value of a cut diamond depends on its color, purity, and size, as well as the skill with which it is cut. In general, the most valuable diamonds are those flawless stones that are colorless or possess a blue-white color. However, the term *blue-white diamond* is sometimes misused by jewelers to describe stones of inferior quality. A faint straw-yellow color that diamonds often possess detracts from the gem's value.

If diamonds are colored deep shades of yellow, red, green, or blue, they are called fancy stones and are greatly prized and command very high prices. Diamonds can be colored deep shades of green by bombardment with nuclear radiation or blue by exposure to high-energy electrons. A stone colored green by irradiation can be heated to bring out a deep yellow hue. These artificially colored stones are difficult to distinguish from natural ones.

Diamonds have been discovered in many localities throughout the world, but only in a few places are they plentiful enough to be mined commercially. Most commonly, diamonds are found in alluvial or placer deposits derived from eroded volcanic mountains. They accumulate in these deposits because of their inert chemical nature, great hardness, and fairly high density. The earliest diamonds were mined from stream gravels in the southern and central portions of India. An estimated 12 million carats (the carat is a unit of weight for gemstones equal to 0.2 gram) was produced from Indian mines. India was virtually the only source of diamonds until they were discovered in Brazil in 1725. East central Brazil has produced about 160,000 carats annually, chiefly from stream gravels near the city of Diamantina, Minas Gerais.

Today, about 95 percent of the world's output of diamonds is from Africa. The Congo is by far the largest producer, supplying more than 50 percent of the global demand. The diamonds are mostly industrial grade, used for cutting tools, and represent only a fraction of the world's total value of diamonds produced. Industrial diamonds are also produced synthetically by sub-

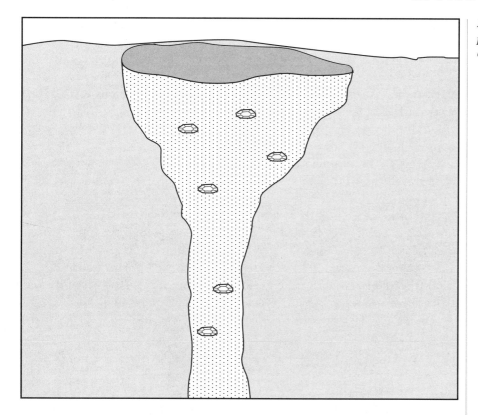

Figure 131 Kimberlite pipes are a major source of diamonds in South Africa.

jecting pure carbon to extreme temperatures and pressures similar to those found deep inside the Earth. Several million carats are manufactured yearly, but these diamonds are not suitable for cutting into gemstones because of their small size.

Although some gem quality diamonds are still recovered from gravels, the principal South African production is from kimberlite pipes (Fig. 131), named for the town of Kimberley, South Africa. They are composed of jumbled fragments of mantle rocks, which are believed to have originated as deep as 150 miles below the surface. The intrusive bodies vary in size and shape, although many are roughly circular and pipe-shaped. Prospecting in South Africa for diamonds has uncovered over 700 kimberlite pipes and other intrusive bodies. However, most of these were found to be barren of diamonds.

The kimberlite deposits were originally worked as open pits, but as the mines became deeper, underground mining methods had to be employed. At the Kimberley mine, the world's deepest diamond mine, the diameter of the pipe at the surface was about 1,000 feet, and the width decreased sharply with depth. Mining stopped in 1908 at a depth of 3,500 feet because of flooding, even though the diamond-rich pipe continued on to greater depths. At the surface of the mine, the kimberlite is weathered to a soft yellow rock, and at depth the rock grades into a harder "blue rock." The ratio of diamonds to barren rock was about 1 to 8 million by weight. The diamonds are extracted from the blue rock by first crushing it finely enough to permit concentration. It is then spread out on tables coated with grease, to which the diamonds adhere while the waste material is washed away.

The world's largest and most productive diamond mine is the Premier mine, located 24 miles east of Pretoria, South Africa. Since mining began in 1903, more than 30 million carats, or about six tons of diamonds, have been produced. The world's largest diamond, the Cullinan, weighing 3,024 carats (21 ounces), was found there in 1905.

Diamonds have been found sparingly in other regions of the world, including Guyana, Venezuela, Australia, and various parts of the United States. Small stones have occasionally been discovered in stream sands along the eastern slope of the Appalachian Mountains from Virginia to Georgia. Diamonds from the gold sands of northern California and southern Oregon have been reported. Sporadic occurrences have also been found along the border of Wyoming and Colorado and in glacial till deposits in Ohio, Wisconsin, and Michigan (Fig. 132).

In 1906, diamonds were discovered in a kimberlite pipe near Murfreesboro, Arkansas. This locality resembled areas containing diamond pipes in South Africa and was the site of the only operating diamond mine in the United States (Fig. 133), yielding a total of about 40,000 stones. It is now a tourist attraction, known as the Crater of Diamonds State Park, where people pay a small fee to sift through the black soil in search of instant wealth. The diamond field is plowed regularly, and generally the period after a rainstorm is the best time to search for diamonds because the surface of the freshly turned soil is washed, exposing the diamonds, which might otherwise look

Figure 132 *The Saukville diamond (left) and Berlington diamond (right) found in Ozaukee and Racine Counties, Michigan, respectively.*

(Photo by W. F. Cannon, courtesy USGS)

Figure 133 *The Arkansas diamond mine south of Murfreesboro, Arkansas, in 1923. A dragline scraper was used to mine diamond-bearing material.*

(Photo by H. D. Miser, courtesy USGS)

like bits of glass. One of the largest diamonds was found by a baby sucking on what was thought to be an ordinary rock.

GOLD AND SILVER

For more than 6,000 years, people have cherished and fought over gold, and this metal was responsible in part for the making of civilization. Gold is the universally accepted medium of exchange. It is also a status symbol, and down through the ages people have adorned their bodies with this metal regardless of its great weight. Along with diamonds, gold is exchanged during marriage ceremonies in most parts of the world. Gold is even believed to have healing powers. People in Japan seek its medicinal powers by bathing in a tub fashioned into a shape of a phoenix made from 400 pounds of pure gold.

Gold does not tarnish and resists corrosion; gold coins recovered from sunken treasure ships that have lain on the bottom of the ocean for centuries look as bright as new. Gold is extremely malleable, and a single ounce can be beaten into a sheet covering nearly 100 square feet. Visitors to Bangkok, Thailand, are often awestruck by the apparent abundance of gold spread on the roofs of temples and other buildings until one realizes how extremely thin gold gilding can be made. Even glass coated with a thin film of gold can reflect the summer's sun and retain a building's heat during the winter to cut down on utility costs.

185

The California gold rush began in early 1848 with the discovery of gold at John Sutter's sawmill near present-day Sacramento. Word spread like wildfire, and Californians headed for the hills to mine gold. They were soon joined by get-rich-quick men from other parts of the country, who stormed into California from all directions. Several thousand poorly equipped fortune hunters died along the way, most from disease, famine, and cold. Mining camps sprang into shantytowns, where miners lived under primitive conditions and claim disputes and drunken brawls were common. Supplies had to be paid for in gold dust, and prices were exorbitant. Of the many thousands who went into the mountains to dig for gold, only a few actually got rich, most of whom did so by mining the pockets of other miners.

The gold-bearing veins of the foothills of the western Sierra Nevada Range in California (Fig. 134) are usually steeply inclined ledges dipping down into the granite roots of the mountains. The hydrothermal veins of the Mother Lode system trend north-south, covering a distance of some 200 miles. The veins are composed of a hard, milky white quartz, generally no more than three feet wide. The quartz might have a few specks of gold and pyrite (fool's gold) sprinkled throughout, but seldom did stringers of pure gold shoot through the veins. Most miners

Figure 134 *Alluvial fans from the Sierra Nevada Range, Death Valley National Monument, Inyo County, California.*

(Photo by H. E. Malde, courtesy USGS)

Figure 135 *Placer gold mining in Borens Gulch, La Plata County, Colorado, in 1875.*

(Photo by W. H. Jackson, courtesy USGS)

panned for gold out of the sands and gravels that washed down from the mountains.

Gold has a specific gravity or density of about 19, making it roughly eight times heavier than ordinary sands and gravels. Therefore, if gold sands are placed in suspension with water by vigorous swirling or sluicing, the gold falls out of the mixture and onto the bottom of a gold pan or sluice box. This technique is known as placer mining (Fig. 135), and for this type of mine to be profitable, many tons of sand and gravel along with large quantities of water have to be processed. An individual panning for gold will most likely not get rich, but at the present price of around $300 a troy ounce (12 ounces per pound), he or she might possibly find enough to pay for provisions with a little left over. The forty-niners rarely did so.

The gold rush was not confined to California but headed eastward into Wyoming, Colorado, New Mexico, and Nevada. The gold of Carlin, Nevada, is an oddity, however. The gold flakes are so small they cannot be seen even with most microscopes. Yet those tiny specks add up to at least 85 million ounces, making the area the world's third-largest gold-producing region. The gold boiled up in the same hot magma plume or hot spot that gave rise to Yellowstone National Park. Along its way to its present location, the hot spot left a trail of basins, volcanoes, and geysers from Nevada, across Idaho, and into Wyoming. About 40 million years ago, the hot spot rested underneath Carlin, where gases and water saturated with gold shot up toward the surface.

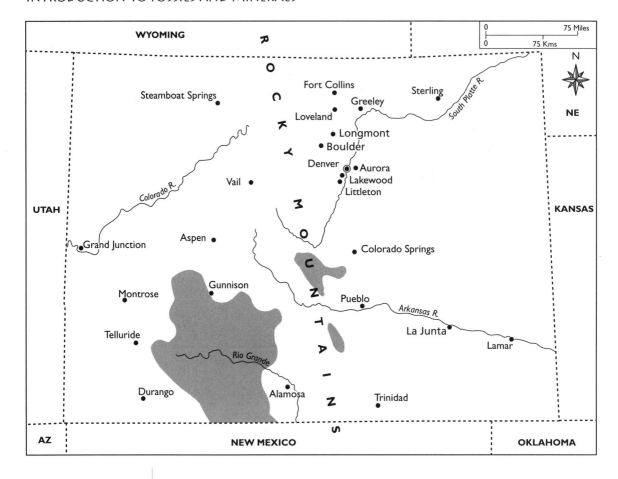

Figure 136 *The stippled areas of the map represent former gold mining districts in Colorado.*

Perhaps the most spectacular mining episodes and mines with the most colorful names were located in the Colorado Mountains (Figs. 136, 137), which are peppered with old abandoned mining camps, attesting to the many thousands who had more gold in their eyes than they found in the ground. When the underground mines played out, gold dredging techniques were developed (Fig. 138). Some of the old mining towns eventually became ski resorts, gambling destinations, and tourist attractions; many more became ghost towns with only a few remnants to mark the past.

Silver is often associated with gold, and the Comstock Lode in Nevada was the scene of one of the largest mining booms in the history of the opening of the American West. Although silver was originally discovered in 1859, production did not peak until the 1870s. Many mines were scattered along a three-mile mineralized fault zone that separated young volcanic rocks from older rocks. The lode forms a slab, inclined about 40 degrees to the horizontal, and reaches a thickness of 400 feet and a depth of 3,000 feet below the surface. The

silver combines with sulfur to make simple silver minerals such as argentite, with a nearly 3 percent gold content, which helped make mining more profitable.

The gold and silver mines in South America were responsible in large part for the Spanish settlement there shortly after Columbus discovered the Americas. Natives of the Inca Empire, which stretched halfway down the Andes Mountains, extending some 3,000 miles from Colombia to Argentina,

Figure 139 *A cross section through the Cerro Rico silver mine, Bolivia, showing the rhyolite porphyry.*

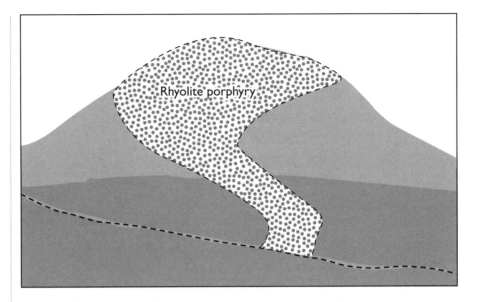

mined gold and silver out of eroded stumps of ancient volcanoes. Cerro Rico ("hill of silver") in Bolivia is a 15,000-foot volcano (Fig. 139) that was literally shot through with veins of silver.

When the Spanish conquistadors first landed in Peru in 1532, they found the Inca Empire torn apart by civil war. The Spaniards had little trouble taking over the empire and captured a great deal of gold and silver along with masterpieces of Inca goldsmiths, which they melted into bullion and sent on to Spain. Some Spanish galleons, heavily loaded with gold and silver, were lost during storms at sea, and today their precious cargoes are eagerly sought after by undersea treasure hunters.

PRECIOUS METALS AND RARE EARTHS

Besides gold and silver, precious metals include those of the platinum group. Platinum is a grayish white heavy metal similar in density to gold, and like gold it is extremely ductile and easily malleable. It is mainly used in jewelry and as a catalyst to accelerate chemical reactions. One notable use of platinum in this manner is in automobile catalytic converters on exhaust systems to oxidize unburned hydrocarbons to further help fight air pollution. As with gold and diamonds, which are resistant to weathering and are extremely dense, causing them to settle to the bottom of sands and gravels, platinum is mined mainly from placer deposits. The best platinum mines are in the Ural Mountains of Russia.

One interesting member of the platinum group is iridium, which is a silver-white, hard, brittle heavy metal that often occurs naturally with platinum. It is relatively abundant on asteroids and comets but extremely rare in the Earth's crust. Iridium along with shock-generated minerals found at suspected meteorite craters are used as evidence of large impacts. Unusually high concentrations of iridium discovered at the boundary between the Cretaceous and Tertiary (K–T) periods (Fig. 140) provide some of the strongest evidence for the impact theory of dinosaur extinction. The boundary layers between other geologic periods associated with mass extinctions contain anomalous amounts of iridium from possible meteorite impacts as well. However, the iridium concentrations are not nearly as strong as those at the end of the Cretaceous, which are as much as 1,000 times background levels, suggesting the K–T event might have been unique in Earth history.

Osmium, another member of the platinum group, is also found on metal-rich asteroids. Osmium is a hard, brittle blue-gray or blue-black metal with an extremely high melting point and is the heaviest known metal. It is used as a catalyst and in manufacturing of hard metal alloys. Iridium and osmium are called siderophiles "iron lovers" because early in the Earth's history they latched onto iron and sank into the planet's interior to form the core;

Figure 140 *The southwest slope of South Table Mountain, Golden, Colorado. The boundary between the Cretaceous and Tertiary Periods lies 10 feet below where the man is standing.*

(Photo by R. W. Brown, courtesy USGS)

that is why they are so rare on the surface. Volcanoes whose magma originates deep within the mantle such as Mauna Loa on Hawaii (Fig. 141) are also sources of significant quantities of iridium and osmium; that characteristic supports the volcanistic theory of dinosaur extinction.

The rare earths are a series of scarce metallic elements with closely similar chemical properties. Rare earths are often found in pegmatites, which are granitic bodies with extremely large crystals. The family of rare earths consists of the lanthanide series of 15 chemical elements, along with scandium and yttrium. Lanthanum is a soft, malleable white metal often used in oil refining. Cerium is a malleable, ductile metal and the most abundant of the rare earths. Praseodymium is a yellowish white metal used chiefly for coloring glass a greenish yellow and is employed in oil refining. Neodymium is a soft yellow metal used in making magnets and lasers and in tinting glass.

Promethium is a radioactive metallic element found as a fission product of uranium, for example, in nuclear power plants. Samarium is a pale gray lustrous metal used in alloys to make permanent magnets. Europium is a soft bivalent and trivalent metal found in monazite sand and used in nuclear research because of its absorption of neutrons. Gadolinium is a magnetic metallic element found in combination with iron, beryllium, and other rare earths. Terbium is a silver-colored trivalent metal; it readily reacts chemically

Figure 141 *Mauna Loa volcano on the main island of Hawaii.*

(Photo courtesy USGS)

with other elements. Dysprosium is a silver-colored metal that forms highly magnetic compounds. Holmium is another metallic element that forms compounds that are highly magnetic.

Erbium, often found with the rare-earth metal yttrium, is a soft metal used in nuclear research because it absorbs neutrons. Thulium is a grayish trivalent metal readily reactive with other chemical elements. Ytterbium is a silver metal that resembles the rare earth yttrium and occurs with it and related elements in several minerals. Lutetium is a metallic element rarely found in nature with no known uses. Scandium is a soft white metallic element that is rare and expensive and has no common industrial uses. Yttrium is a grayish metal used for making permanent magnets, lasers, and superconducting materials.

Certain types of rare earths are important in making superconductors. These materials are capable of conducting electrical currents at near-zero resistance at very low temperatures. At temperatures approaching absolute zero (-273 degrees Celsius), the electrical resistance in a conducting medium decreases to essentially zero, so that currents can flow in electromagnets for a long time before dying down. Ceramics doped with rare earths, which give them metallike properties, have greatly increased the temperature of superconducting materials. Perhaps the discovery of new materials for superconductivity at higher temperatures will revolutionize the way we live in the future.

The next chapter will discuss the rare and unusual rock types, which add extra excitement to the world of geology.

9

THE RARE AND UNUSUAL

ROCKS WITH UNIQUE PROPERTIES

E very once in a while, someone makes a momentous discovery that defies preconceived notions about the "real world." Physicists study the microworld and believe that by breaking atoms into tiny pieces they can tell us something about the origin of the universe. Geologists study the macroworld in search of clues about how the Earth originated. Astronomers study the megaworld and look to the stars to tell us something about our galaxy and the vast expanse that lies beyond.

Meanwhile, plenty of unsolved riddles await us down here on Earth. For instance, what was the purpose of the megaliths at Stonehenge in southern England (Fig. 142)? Was the volcanic island of Thera, which exploded in 1580 B.C., really the fabled continent of Atlantis? How did Ayers Rock, the largest rock in the world, measuring 5.5 miles long and 1,100 feet high, happen to appear in the middle of the desert in central Australia? The rocks will ultimately help solve some of our most puzzling questions and might some day tell us why the dinosaurs disappeared.

Figure 142 Megalithic monuments scattered throughout Europe might have had an astronomical function.

ROCKS THAT FOLLOW THE SUN

Hundreds of millions of years ago, single-celled organisms recorded in rock the interactions of the Sun, Earth, and the Moon. Ancestors of blue-green algae built concentrically layered mounds resembling cabbage heads called stromatolite structures by cementing sediment grains together with a gluelike substance secreted from their bodies (Fig. 143). As with modern stromatolites, the ancient stromatolite colonies grew in the intertidal zone, and their height was indicative of the height of the tides during their lifetime. This is because stromatolites grow between the low-tide mark and the high-tide mark.

The stromatolite structures also tilted toward the Sun as they grew. This phenomenon is known as heliotropism (*helio* meaning "sun"), which is the inclination of a structure toward the average direction of sunlight. Stromatolites found in the Bitter Springs Formation in central Australia provided an 850-million-year-old fossil record of the Sun's movement across the sky. A stromatolite situated near the equator pointed south in the winter and north in the summer and developed a growth pattern in the shape of a sine wave.

If new sediment layers were constructed each day, then the number of layers appearing in one wavelength represented the number of days in a year during the time of the stromatolite's growth. By counting the layers in stromatolite fossils, researchers have estimated that approximately 435 days constituted a year (the time required for the Earth to complete one revolution around the Sun) during the late Proterozoic. The results agreed well with

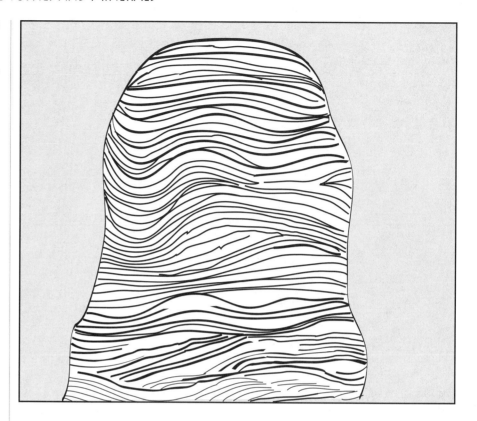

Figure 143 *Stromatolites are layered structures formed by colonies of primitive blue-green algae.*

those of counting the growth rings of ancient coral fossils to estimate the number of days in a year going as far back as the beginning of the Cambrian period, 570 million years ago. The studies indicated that the Earth was spinning faster on its axis and that the days were only 20 hours long.

In addition, the sine wave patterns of ancient stromatolites contained information about the maximum travel of the Sun across the equator. The equator forms an oblique angle to the plane of Earth's orbit around the Sun, called the ecliptic. This angle is controlled by the tilt of the Earth's rotational axis. The maximum latitude of the Sun during the peak of each season is obtained by measuring the maximum angle at which the sine wave deviates from the average direction of stromatolite growth. Today, the Sun travels 23.5 degrees north of the equator during the summer and 23.5 degrees south of the equator during the winter. However, about 850 million years ago, this value was about 26.5 degrees, an indication that the climate at that time would have been much more seasonal than it is today. This observation supports the idea that the axial tilt angle has been decreasing with time.

Present-day stromatolites live in the intertidal zones, above the low-tide mark, and their height is indicative of the height of the tides, which is mostly controlled by the gravitational pull of the Moon. The stromatolite colonies of the Warrawoona group in North Pole, Western Australia, at 3.5 billion years, the oldest on Earth, grew to tremendous heights with some over 30 feet tall. This suggests that an early age the Moon was much closer to the Earth, and because of its stronger gravitational attraction at this range, it raised tremendous tides that must have flooded coastal areas several miles inland.

The data also explain why the length of day was much shorter. The early Earth rotated much faster than it does today, and as its rotational rate gradually slowed as a result of drag forces produced by the tides, some of its angular momentum (rotational energy) was transferred to the Moon, flinging it out into a wider orbit. Even today, the Moon is receding from the Earth at a rate of about two inches per year.

The solar cycle is a periodic fluctuation of solar output, occurring today about every 22 years, double the 11-year sunspot cycle. Evidence for a solar cycle operating as far back as the Precambrian is thought to exist in 680-million-year-old glacial varves, or banded deposits, found in lake bed sediments north of Adelaide in South Australia. The varves consist of alternating layers of silt deposited annually during the late Precambrian ice age (Fig. 144). Each summer when the glacial ice melted, sediment-laden meltwater discharged into a lake below the glacier, and the sediments settled out to form a stratified deposit (Fig. 145).

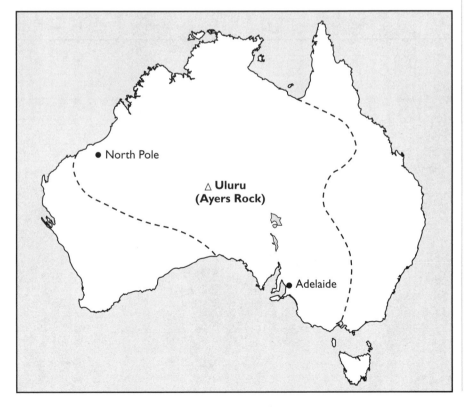

Figure 144 *The dashed lines indicate the extent of the late Precambrian glaciation in Australia.*

Figure 145 *Australian desert varves comprised cyclic laminations of sandstones and siltstones.*

During times of intense solar activity, the Earth's climatic temperature increased slightly, causing more glacial melting and the deposition of thicker varves. By counting the layers of thick and thin varves, scientists can establish a stratigraphic sequence that mimics the 11-year sunspot cycle, the occurrence of large numbers of sunspots on the Sun's surface, and the 22-year solar cycle or possibly even the early lunar cycle, covering the full range of lunar orbital variations, which today is about 19 years.

Another type of stratified deposit, which might well be the most beautiful, valuable, and enigmatic rock ever created on the planet, occurs in banded iron formations. Composed of alternating layers of iron and silica, they were formed about 2 billion years ago at the height of the earliest ice age. For unknown reasons, major episodes of iron deposition coincided with periods of glaciation. Evidently, on average the oceans were much warmer at that time than they are at present. When iron- and silicate-rich warm currents flowed toward the glaciated polar regions, the suddenly cooled waters could no longer hold the minerals in solution. They thus precipitated out, forming alternating layers due to difference in settling rates between silica and iron, the heavier of the two minerals.

Volcanic eruptions are also known to follow the 11-year solar cycle, which is a slight waxing and waning in the sun's energy output. The study of hundreds of eruptions over the past four centuries implies that the solar cycle might have had an influence on the time when volcanoes came to life. The eruptions appeared most numerous during the weakest portion of the solar cycle, when the number of sunspots are low. During the peak of the solar cycle, emissions from the Sun cause small but abrupt changes in the Earth's atmosphere, jarring the planet slightly. This motion might trigger tiny earthquakes that relieve stress under volcanoes, thereby preventing a large eruption until the solar cycle is again at a minimum, when the pent-up volcanoes blow their tops.

MINERALS THAT REVERSE THEMSELVES

Over the last 170 million years, for still unexplained reasons, the Earth's magnetic field has reversed polarity some 300 times; the last reversal occurred about 780,000 years ago. During a reversal, the Earth temporarily lets down its magnetic shield, which protects it from cosmic radiation from outer space. Proof that the magnetic poles periodically change places is found on the ocean floor near spreading centers, where new oceanic crust is being generated. As layers of basalt cool, they become slightly magnetized and acquire the polarity of the magnetic poles at the time of their deposition.

Furthermore, one set of alternating magnetic bands of basalt is the mirror image of the opposite set on the other side of the spreading ridge (Fig. 146). This property became the conclusive proof for seafloor spreading because in order for the magnetic strips to form in such a manner the seafloor had to be spreading apart from a common center. In addition, the magnetic stripes provide a means of dating virtually the entire ocean floor because the magnetic reversals occur randomly, and any set of patterns is unique in Earth history.

Magnetite, which faithfully records the Earth's magnetic field, was thought to be the dominant magnetic mineral in rocks. But in the early 1950s, a rare mineral called titanohematite, composed of iron, titanium, and oxygen, was found to have the odd ability to become magnetized in the opposite direction from the Earth's magnetic field. This unusual behavior could have played havoc with scientists trying to prove the theory of magnetic pole reversal. Their existence could also complicate the analysis of the magnetic orientations in rocks used to date lava flows.

Although once thought to be exceedingly rare, self-reversing minerals more recently have been found in sedimentary basins and lava fields of western North America (Fig. 147), where, in some places, such as the Bighorn Basin of Wyoming and the San Juan Basin in Mexico, they are the dominant magnetic minerals. Relatively abundant titanohematites were also found in 10,000-year-old lava flows of California's Mount Shasta volcano (Fig. 148). The mineral is associated with explosive eruptions common in the Cascade

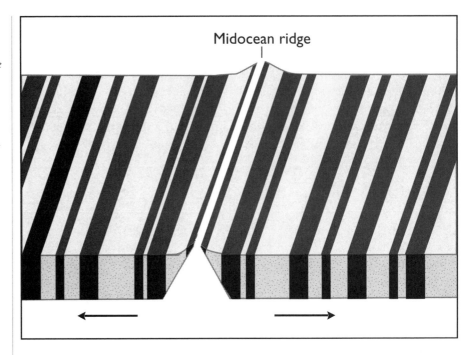

Figure 146 *Magnetic stripes on the ocean floor are mirror images of each other and indicate that the ocean crust is spreading apart.*

Midocean ridge

Range, as witnessed by the huge lateral blast of Mount Saint Helens in 1980. Because self-reversing titanohematites form under special conditions, their presence in volcanic rock might help volcanologists understand the nature of magma that rises up through volcanoes.

Normally, when rocks are imprinted with a magnetic field, the magnetic fields of their atoms line up with the Earth's magnetic field as they cool past their Curie point, the temperature at which a magnetic field becomes permanent. Below that point, the field freezes until such a time as the rock is reheated, destroying the magnetic field. Thus, the magnetic fields can be thought of as tiny fossil compasses, pointing in whatever direction the Earth's magnetic field happened to be during deposition. Self-reversing minerals, however, have two Curie points, one occurring at a higher temperature than the other. As the mineral cools, it develops two magnetic regions. The first has its magnetic field aligned with the Earth's, and the second has a much stronger field aligned in the opposite direction. Thus, the final polarity of the rock's magnetic field is reversed.

HALOS OF STONE

In the permafrost regions of the Arctic, soil and rocks are fashioned into strikingly beautiful and orderly patterns that have confronted geologists for

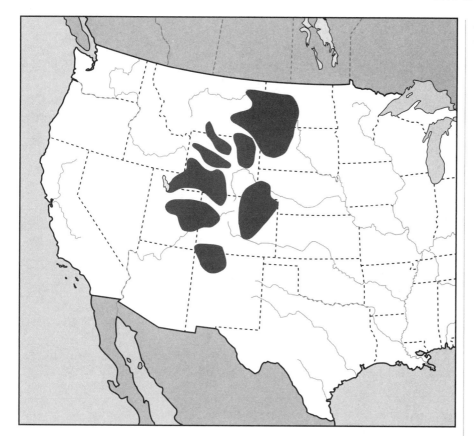

Figure 147 *A map indicating the location of self-reversing minerals in the western United States.*

centuries. Every summer, the retreating snows unveil a bizarre assortment of rocks arranged in a honeycomblike network as the ground begins to thaw, giving the landscape the appearance of a tiled floor. These patterns are found in most of the northern lands and alpine regions, where the soil is

Figure 148 *Mount Shasta in the Cascade Range, Siskiyou County, California.*

(Photo by C. D. Miller, courtesy USGS).

Figure 149 *Stone polygons east of Maclaren River, Clearwater Mountains, Valdez Creek District, Alaska. Note the complete sorting between the clay-rich centers and the bordering lines of coarse boulders.*

(Photo by C. Wahrhaftig, courtesy USGS)

exposed to moisture and seasonal freezing and thawing. The polygons range in size from a few inches across when composed of small pebbles to several tens of feet when large boulders form protective rings around mounds of soil (Fig. 149).

The polygons were probably produced by forces similar to those that cause frost heaving, which pushes rocks up through the soil. This phenomenon is well known to northern farmers, as every spring produces a new crop of stones in their fields. The boulders move through the soil either by a pull from above or by a push from below. If the top of the rock freezes first, it is pulled up by the expanding frozen soil. When the soil thaws, sediment gathers below the rock, which settles at a slightly higher level. The expanding frozen soil lying below could also heave the rock upward. After several freeze-thaw cycles, the boulder finally rests on the surface. Rocks also have been known to push through highway pavement, and fence posts have been shoved completely out of the ground by frost heaving.

The regular polygon patterned ground might have been formed by the movement of soil of mixed composition upward toward the center of the mound and downward under the boulders. The material moves down under the gravel in an action similar to that in a subduction zone. The soil might move in convective cells (Fig. 150), much as bubbles rise up in a pot of boiling water. The coarser material, composed of gravel and boulders, is gradually shoved radially outward from the central area and subducted along the edges, leaving the finer materials behind. This idea is supported by the fact that the material in the center of the mound often appears churned up. The opposite condition occurs during the formation of stone pits, which are in every respect the inverse of stone polygons. Instead of convection cells circulating up through the center of the polygon and down along the borders, in stone pits the circulation is down through the center and up along the outside.

The arctic soil provides an assortment of other geometric designs, including steps, stripes, and nets that lie between the circles and polygons. These forms can reach 150 feet in diameter. Relics of ancient surface patterns shaped during the last ice age measuring up to 500 feet have been found in former permafrost regions. Even images of Mars sent back from spacecraft depict furrowed rings, polygonal fractures, and ground-ice patterns of every description, suggesting that the planet's surface water has long since gone underground.

Other examples of patterned ground include polygonal shapes created in desert muds (Fig. 151), which were formed by the contraction of the mud when it dried rapidly in the hot Sun. Sorted circles also might form by the increased wearing down of coarse grains in isolated cracks in bedrock. Even vibrations from earthquakes are thought to cause the sorting of some sediments. Pebbles embedded in sand rise toward the surface during ground shaking as a result of convective motions in the sediment. Vibrating sands at the surface congregate into stable patterns, including hexagons, squares, stripes, and circular heaps. Also, curious occurrences of patterns form in sediments on the ocean floor.

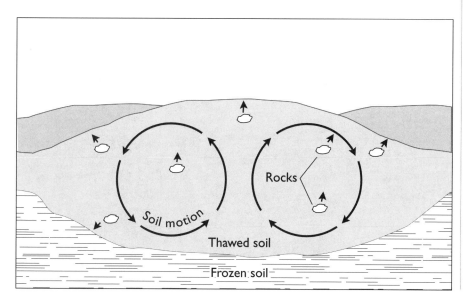

Figure 150 *The formation of polygonal structures by convective cells, which move coarser material to the surface.*

WHISTLING STONES

If a volcano erupts explosively and casts fluid lava high into the air, the lava is dispersed by the wind, giving rise to particles of various sizes from ash to molten blobs of lava up to 15 feet wide called volcanic bombs (Fig. 152). Often volcanic bombs are still soft enough to change shape during their flight through the air. Their motion in flight tends to give them elongated shapes and smooth surfaces. Rounded masses are common, and sometimes the ends are twisted into a spiral. Volcanic bombs frequently flatten or splatter when they land. If the bombs are the size of a nut, they are called lapilli, Latin for "little stones," and form strange gravellike deposits that litter the countryside.

If volcanic bombs cool in flight, they form a variety of shapes. The various types of volcanic bombs are described as cannonball, spindle, bread-crust, cow dung, ribbon, or fusiform, depending on their shape or surface appearance. Bread-crust bombs, which often reach several feet across, are named for

Figure 151 *Mud cracks on the tidal flats of San Francisco Bay, California.*

(Photo by D. H. Radbruch-Hail, courtesy USGS)

Figure 152 *Volcanic bombs at the southwest base of a cinder cone, 10 miles northwest of Lassen Peak, Cascade Range, Shasta County, California.*

(Photo by J. S. Diller, courtesy USGS)

their crusty appearance, which is caused by gases escaping from the bomb while the outer surface is hardening. The bombs might also be hollow or have numerous cavities formed by bursting gas bubbles on the surface. Sometimes volcanic bombs actually explode when they hit the ground as a result of the rapid expansion of gas in the molten interior when the solid crust cracks open on impact. Often, volcanic bombs spin wildly through the air, causing them to whistle like incoming cannon fire.

Beneath the Pacific Ocean near French Polynesia, strange single-frequency notes were found emanating from clouds of bubbles billowing out of undersea volcanoes. The notes were among the purest in the world, far better than those played by any musical instrument. The sounds were at first thought to have come from secret military testing or perhaps some strange marine creature. Even whales are known to emit loud calls to one another. However, no explosion or animal was capable of sounds of such singular notes. The low frequency of the sound also meant the source had to be quite large. Further search of the ocean depths uncovered a huge swarm of bubbles. When undersea volcanoes gush out magma and scalding water, the surrounding water boils away into bubbles of steam. As the closely packed bubbles rise toward the surface, they rapidly change shape, producing extraordinary single-frequency sound waves.

In the desert, a curious feature of sand dunes is an unexplained phenomenon known as booming sands. The sound occurs almost exclusively in large, isolated dunes deep in the desert or well inland from the coast. The noises can be triggered by simply walking along the dune ridges. When sand slides down the lee side of a dune, it sometimes emits a loud rumble. The sounds emitting from the dunes have been likened to bells, trumpets, pipe organs, foghorns, cannon fire, thunder, buzzing telephone wires, and low-flying aircraft. The grains found in sound-producing sand are usually spherical, well rounded, and well sorted, or of equal size. The sound appears to originate from some sort of harmonic event occurring at the same frequency. However, normal landsliding involves a mass of randomly moving sand grains that collide with a frequency much too high to produce such a peculiar noise.

STONE PILLARS

Sandstone pillars stretching as much as 190 feet above the desert floor near the town of Gallup in northwestern New Mexico appear to be the world's largest trace fossils, as they are composed of more than 100 fossilized termite nests. Trace fossils are the fossilized tracks, trails, borings, or burrows formed by ancient organisms. The termite nests are around 155 million years old and are the first that have been found from the Jurassic period. Similar fossilized termite nests dating to 220 million years old have been discovered in Arizona's Petrified Forest National Park. Previously, social insects, such as termites, ants, and bees, were thought not to exist until about 100 million years ago when flowering plants first appeared. Interestingly, entomologists, biologists who study insects, have known for some time that termites probably lived early in the triassic period, just about the time the dinosaurs came onto the scene.

On land, tufa is a porous rock (the compact, dense variety is called travertine) that commonly occurs as an incrustation around the mouths of hot springs and are composed of calcite or silica. But in southwestern Greenland, more than 500 giant towers of tufa cluster together in the chilly waters of Ikka Fjord. Some reach as high as 60 feet, and their tops are visible at low tide. The towers are made of an unusual form of calcium carbonate called ikaite, whose crystals form when carbonate-rich water from springs beneath the fjord seeps upward and comes into contact with cold, calcium-laden seawater. Because of the low temperature, the water cannot escape during the precipitation of the mineral and is incorporated into the crystal lattice, producing weird, yet beautiful formations. The towers also provide shelter for algae, anemones, and sea cucumbers.

Another undersea curiosity are huge pillars of lava standing like Greek columns on the ocean floor as much as 45 feet tall. How these strange spires formed remains unclear. The best explanation suggests that the pillars were created by the slow advances of lava oozing from volcanic ridges. Several blobs of lava nestle together in a ring, leaving an empty water-filled space in the center. The sides of these adjoining blobs form the pillar walls, as the outer layers cool on contact with seawater. The insides of the blobs remain fluid until the lava flows back into the vent, as though the drain plug were pulled. The fragile blobs then collapse, resembling large empty eggshells, leaving hollow pillars formed from the interior walls of the ring behind.

Forests of exquisite chimneys up to 30 feet tall called black smokers (Fig. 153) spew hot water blackened with sulfide minerals into the near-freezing deep abyss. The vents have openings typically ranging from less than a half-inch to more than six feet across. They are common throughout the world's oceans along the midocean spreading ridge system and are believed to be the main route through which the Earth's interior loses heat. The vents exhibit a strange phenomenon of glowing in the pitch-black dark, possibly as a result of the sudden cooling of the 350° Celsius water, which produces crystalloluminescence as dissolved minerals crystallize and drop out of solution, thereby emitting light. The light, although extremely dim, is apparently bright enough to allow photosynthesis to take place even on the very bottom of the deep sea.

The ghostly white chimneys of the Mariana seamounts in the western Pacific near the world's deepest trench are composed of a form of aragonite, a white calcium carbonate rock with a very unusual texture that normally dissolves in seawater at these great depths. Hundreds of corroded and dead carbonate chimneys were strewn across the ocean floor in wide "graveyards." The

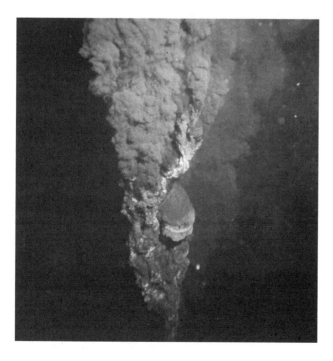

Figure 153 A black smoker on the East Pacific Rise, showing high-temperature water blackened with sulfide minerals spewing out of the vent.

(Photo by R. D. Ballard, courtesy WHOI)

fluid temperatures in subduction zones are cool compared to those associated with midocean ridges, enabling crystals of aragonite and calcite to form. Apparently, cool water slowly seeping from beneath the surface allows the carbonate chimneys to grow and avoid dissolution by seawater. Many carbonate chimneys are thin and generally less than six feet high. Other chimney structures are thicker and much taller and occasionally cluster together into ramparts.

MINERALS THAT GLOW IN THE DARK

When rays from an ultraviolet lamp called a black light strike certain minerals, their atoms become excited (rapidly vibrate) and reradiate energy at a lower wavelength, often providing strikingly brilliant colors. This effect is termed fluorescence, named after the mineral fluorite, which often glows blue-violet when a black light is shone on it. Fluorescence is the ability of an object to absorb energy such as ultraviolet light and reradiate that energy at a different wavelength, usually in the visible range.

Ultraviolet light has a very short wavelength (Fig. 154) that cannot be seen by the naked eye. However, some minerals when exposed to a black light emit light at a longer wavelength that is visible only as long as the light source remains. A good source of ultraviolet light of short wavelengths is a quartz lamp, whereas an argon light produces longer ultraviolet rays. Other sources of fluorescent energy are X rays and cathode rays, which are high-energy electrons.

The overwhelming majority of fluorescent minerals are not very attractive until a black light is turned on them, when the ultraviolet rays produce vivid and intense colors. The distinctive colors are produced by a variety of minerals. Some minerals might fluoresce in one locality but not in another. The property of fluorescence aids the geologist in prospecting for minerals and petroleum at the surface and the hobbyist in discovering this exciting field of mineralogy.

Figure 154 *The electromagnetic spectrum showing ultraviolet light above the visible range.*

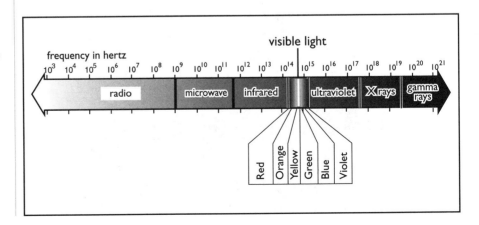

Fluorescent minerals usually contain a small amount of impurities such as manganese, which are called activators. These enhance a mineral's ability to glow in the presence of ultraviolet rays. If a mineral continues to glow for a short time (up to several minutes) after the ultraviolet light source has been removed, it is phosphorescent. These minerals are much less common than those that fluoresce. In addition, some minerals only fluoresce in shortwave ultraviolet, whereas others only fluoresce in long-wave ultraviolet. However, many fluorescent minerals glow in both ultraviolet wavelengths. Therefore, when exploring for fluorescent minerals, one is advised to select a black light that uses both ultraviolet wavelengths.

Willemite, which is a zinc ore, was first discovered to have fluorescent properties at the Franklin mine in New Jersey, where spectacular fluorescent specimens are perhaps the finest in the world. Miners observed a brilliant green glow from willemite when they exposed it to the ultraviolet rays of an arc lamp in a dark mine shaft. Franklin calcite also fluoresces, providing a brilliant red color. Sphalerite, another ore of zinc, is also fluorescent and offers another quality by emitting flashes of light when scratched in the dark.

Scheelite, the most important tungsten ore in North America, is one of the few minerals that always fluoresce. Geologists explore for scheelite by shining a black light on suspected outcrops at night and look for a characteristic blue glow. If the mineral glows yellow, molybdenum, an important mineral for hardening steel, is present. Scapolite, a complex silicate metamorphic mineral, glows an appealing yellow under a black light. Opal from the western states glows green as a result of the presence of a small amount of uranium impurities. Uranium ores, besides being detected by their radioactivity, produce strikingly beautiful hues of yellow and green under ultraviolet light. Even diamonds will fluoresce under a black light with various hues of blue.

LIGHTNING FAST GLASS

When two boys stumbled on an unusual rock near Winans Lake, Michigan, they thought they had found what appeared to be at first glance a huge dinosaur leg bone. Scientists from the Museum of Paleontology at the University of Michigan went out to investigate, only to find that the 15-foot-long white, green, and gray object was the world's largest fulgurite, Latin for "thunderbolt." It is a tube-shaped glob of glass that formed when a powerful lightning bolt struck the ground (Fig. 155). The glassy tubes of lightning-fused rock are most common on mountaintops, probably because their high altitude attracts more lightning strikes. Although fulgurites might form from any type of rock, the largest have been created from unconsolidated (loose) sand.

For centuries, scientists have known that large lightning bolts, which can attain temperatures several times those on the surface of the Sun, can melt or vaporize rocks they strike. However, until recently, the chemical and physical processes in the formation of fulgurites were largely unknown. In addition, the studies revealed the presence of two metallic minerals that had never before been found to occur in nature. The fulgurite also was found to be one of the most chemically reduced (oxygen-removed) natural substances known on Earth.

Electron microscope analysis of metallic globules embedded in the fulgurite glass showed them to be composed of a variety of iron and silicon metal compounds previously known only to exist in meteorites. Apparently, the lightning bolt somehow chemically altered the original iron oxides in the ground, to an even greater extent than those found in most meteorites. The fulgurite was also enriched in gold, which the lightning apparently scavenged from the surrounding soil and concentrated in the glass.

A glass of a different sort is in the form of tektites (Fig. 156); the name is derived from the Greek *tektos,* meaning "molten." They are glassy bodies created from the melt of a large meteorite impact. Over half the rock ejected by an impact remains molten in the rising plume and falls back to Earth as tektites. Much of the high-flying material is known to be deposited halfway

Figure 155 *A large lightning bolt.*

(Photo courtesy National Oceanic and Atmospheric Administration [NOAA])

Figure 156 *A North American tektite found in Texas in November 1985, showing surface erosional and corrosional features.*

(Photo by E.C.T. Chao, courtesy USGS)

around the world. Massive meteorite impacts dump millions of tons of tektites over vast areas called strewn fields.

Tektites range in color from bottle-green to yellow-brown to black; they were once prized as ornaments by the Cro-Magnon, our ancient human ancestors. Tektites are usually small, about pebble-sized, although a few have been known to be as large as cobbles. Tektites are chemically distinct from meteorites and have a composition similar to that of the volcanic glass obsidian but contain much less gas and water. They also lack microcrystals, a characteristic uncommon in any kind of volcanic glass.

Tektites comprise abundant silica similar to the pure quartz sands used to manufacture glass. Indeed, tektites appear to be natural glasses formed by the intense heat generated by a large meteorite impact. The impact flings molten material far and wide; while airborne, the liquid drops of rock solidify into various shapes from irregular to spherical, including ellipsoidal, barrel, pear, dumbbell, and button shapes. They also have distinct surface markings that apparently formed while solidifying during their flight through the air.

Mysterious glass fragments strewn over Egypt's Western Desert appear to be melts from a huge impact about 30 million years ago. Large, fist-size, clear glass fragments found scattered across the Libyan Desert were analyzed for rare trace elements, which indicated that the glasses were produced by an impact into the desert sands. The large fragments of glass showed extraordinary clarity. The impact force also generates extremely high temperatures that fuse sediment into small glassy spherules resembling volcanic glass. Thick deposits of sand-size spherules scattered throughout the world document the great meteorite bombardments during the Earth's entire history.

STONES FROM THE SKY

The origin of meteorites, which are metallic or stony bodies that enter the Earth's atmosphere and impact on the ground (Fig. 157), has been a longstanding puzzle. The most accepted theory is that they are from the asteroid belt lying between the orbits of Mars and Jupiter. Most meteorites showering down on Earth arrive from the main asteroid belt, a 250-million-mile-wide band of primordial debris. The asteroid belt contains bodies ranging in size from small grains called micrometeorites to huge chunks of rock upward of hundreds of miles wide called asteroids.

When asteroids collide, the impacts chip away at their surfaces, providing numerous small fragments that often fall to Earth as meteorites. Igneous asteroids, called S types, are most frequently found in the inner region of the asteroid belt and are apparently the source of the most common class of meteorites, known as the ordinary chondrites. Certain rare meteorites might be pieces of the lunar or Martian crust blasted out by large asteroid impacts.

Meteorite falls are more common than most people realize. Every day thousands of meteorites rain down onto the Earth, and occasional meteor showers can involve hundreds of thousands of stones. Nearly 1 million tons of meteoritic material is produced annually. Most meteors completely burn up on entering the atmosphere, and their ashes contribute to the load of atmospheric dust, which is largely responsible for our blue skies and red sunsets. The remainder that survive the blazing journey through the atmosphere can cause havoc, as numerous examples of meteorites crashing into houses and automobiles attest.

Meteorites have been observed throughout human history. Historians have often argued that a spectacular meteorite fall of 3,000 stones at l'Aigle in the French province of Normandy in 1803 sparked the early investigation of meteorites. Yet this spectacle was actually eclipsed nine years earlier by a massive meteorite shower in Siena, Italy, on June 16, 1794. It was the most significant fall in recent times and gave birth to the modern science of meteoritics.

The ancient Chinese were perhaps the earliest to report falling meteorites, during the seventh century B.C. An interesting note is that Chinese meteorites are rare, and to date no large impact craters have been recognized in China. The first report of a meteorite impact on the Moon was a flash wit-

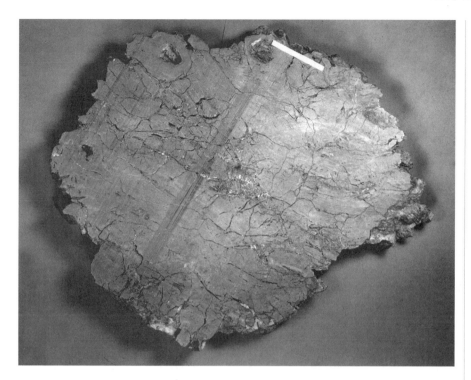

Figure 157 *The Wolf Creek meteorite from Western Australia, showing crack development on cut surface.*

(Photo by G.T. Faust, courtesy USGS)

nessed by a Canterbury monk on June 25, 1178. The Moon has been bombarded by enough small asteroids to account for every lunar crater less than a mile across formed during the last 3 billion years.

The oldest meteorite fall of which material is still preserved in a museum is a 120-pound stone that landed outside Ensisheim in Alsace, France, on November 16, 1492. The largest meteorite found in the United States is the 16-ton Willamette Meteorite, which crashed to Earth sometime during the past million years. Discovered in 1902 near Portland, Oregon, it measured 10 feet long, seven feet wide, and four feet high.

One of the largest meteorites actually seen falling from the sky was an 880-pound stone that landed in a farmer's field near Paragould, Arkansas, on March 27, 1886. The largest known meteorite find, named Hoba West, was located on a farm near Grootfontein, South-West Africa (Namibia), in 1920 and weighs about 60 tons. The heaviest observable stone meteorite landed in a cornfield in Norton County, Kansas, on March 18, 1948. It dug a pit in the ground three feet wide and 10 feet deep. A 40-pound meteorite that landed in Nigeria, Africa, in 1962 was identified as having been a piece of Mars ejected by a massive collision millions of years ago.

Over 500 major meteorite falls strike yearly, most of which plunge into the ocean and accumulate on the seafloor. The braking action of the atmosphere slows the entry of the great majority of meteorites that land on the surface, so they only bury themselves a short distance into the ground. Not all meteorites are hot when they land because the lower atmosphere tends to cool the rocks, which, in some cases, are covered by a thin layer of frost.

The most easily recognizable meteorites are the iron variety, although they only represent about 5 percent of all meteorite falls. They are composed principally of iron and nickel along with sulfur, carbon, and traces of other elements. Their composition is thought to be similar to that of the Earth's metallic core and might once have comprised the cores of large planetoids that disintegrated eons ago. Because of their dense structure, iron meteorites tend to survive impacts intact, and most are found by farmers plowing their fields.

The most common type of meteorite is the stony variety, which constitutes some 90 percent of all falls. But because they are similar to Earth materials and therefore erode easily, they are often difficult to find. The meteorites are composed of tiny spheres of silicate minerals in a fine-grained rocky matrix called chondrules, from the Greek *chondros,* meaning "grain." Chondrules are believed to have formed from clumps of precursor particles when the solar system was emerging from a swirling disk of gas and dust, and the meteorites that contain them are known as chondrites.

Perhaps the world's largest source of meteorites is the Nullarbor Plain, an area of limestone that stretches 400 miles along the south coast of Western and South Australia. The pale, smooth desert plain provides a perfect backdrop for spotting meteorites, which are usually colored dark brown or black. Since the desert experiences very little erosion, the meteorites are well preserved and found just where they land. Over 1,000 fragments have so far been recovered from 150 meteorites that fell during the last 20,000 years. One extremely large iron stone, called the Mundrabilla Meteorite, weighed more than 11 tons.

One of the best hunting grounds for meteorites happens to be on the glaciers of Antarctica. Some of the meteorites landing on Antarctica are believed to have come from the Moon and even as far away as Mars (Fig. 158). A meteorite from the Allan Hills region of Antarctica was composed of diogenite, a common type of basalt from the asteroid belt, possibly impact-blasted out of the crust of Mars and hurled toward Earth. Organic compounds found on a Martian meteorite landing in Antarctica hint of previous life on Mars. It wandered in space for some 3 million years before finally being captured by the Earth's gravity.

Currently, numerous large circular structures are spread around the world, possibly resulting from substantial meteorite impacts. One of the largest impact structures is outlined by the distinctively circular Manicouagan Reservoir (Fig. 159) in east central Quebec, which is nearly 60 miles in diameter. The New Quebec Crater in northern Canada is two miles in diameter, 1,300 feet deep, and filled with water to form one of the world's deepest lakes. The best preserved meteorite impact crater is Meteor Crater (Fig. 160) in the

Figure 158 A meteorite discovered in Antarctica believed to be of Martian origin.

(Photo courtesy National Aeronautics and Space Administration [NASA])

Arizona desert near Winslow; it measures about 4,000 feet across and several hundred feet deep. It was gouged out roughly 50,000 years ago by a meteorite weighing over 60,000 tons.

Figure 159 The Manicouagan impact structure, Quebec, Canada.

(Photo courtesy NASA)

Figure 160 *Meteor Crater near Winslow, Arizona.*

(Photo courtesy USGS)

When a large meteorite slams into the Earth, it kicks up a great deal of sediment. The finer material is lofted high into the atmosphere, and the coarse debris falls back around the perimeter of the crater, forming a high, steep-banked rim. Not only are the rocks shattered in the vicinity of the impact, but the shock wave also causes shock metamorphism of the surrounding rocks, changing their composition and crystal structure. The most readily recognizable shock effect is the fracturing of rocks into distinct conical and striated patterns called shatter cones. They form most readily in fine-grained rocks that have little internal structure, such as limestone and quartzite.

Large meteorite impacts also produce shocked quartz grains that are characterized by prominent striations across crystal faces. Minerals such as quartz and feldspar develop these features when high-pressure shock waves exert shearing forces on the crystals, producing parallel fracture planes called lamellae. The presence of shocked quartz in sedimentary deposits at the Cretaceous-Tertiary (K-T) boundary all around the world is considered evidence that the Earth was struck by a large meteorite that possibly ended the reign of the dinosaurs.

The concluding chapter will explain where and how dinosaur fossils as well as other fossils and minerals are found.

10

WHERE FOSSILS AND MINERALS ARE FOUND

F ossils and rock specimens can be found by the amateur nearly every-
where. The vast majority of fossils are located in ancient marine sediments, some of which accumulated when inland seas invaded the
continents during times of rising sea levels. Most minerals found in sedimentary rocks were precipitated from seawater. When land is eroded, some
3 billion tons of rock is dissolved by water and carried by streams to the
sea each year. This is sufficient to lower the entire land surface of the Earth
by as much as an inch in 2,000 years. It is also one of the reasons why the
ocean is so salty. Besides ordinary table salt, seawater contains large amounts
of calcium carbonate, calcium sulfate, and silica. These minerals precipitate
from seawater by biologic or chemical processes. They also can replace
other minerals or the skeletal remains in fossils.

Many fossils can be found in abandoned limestone quarries and gravel
pits, where the rocks are well exposed and conveniently broken up. Also,
abundant fossil plant leaves and stems can be found in abandoned coal pits.
Because many old abandoned ore mines are excavated in igneous rocks, in
which most minerals are found, these places as well as other granitic outcrops

Figure 161 Sandstone cliffs of late Cretaceous age in Mesa Verde National Park, Colorado.

(Photo courtesy National Park Service)

are often excellent sites for finding minerals with good crystals. In addition to these, rock outcrops (Fig. 161), road cuts, stream beds, and sea cliffs offer good rock exposures for collecting fossils and minerals.

FOSSIL-BEARING ROCKS

In areas where marine sediments outcrop, chances are good that these sediments are fossiliferous, meaning they contain abundant fossils. Indeed, few places in the United States lack fossils, because most parts of the North American continent have been invaded by seas at various times in the past, allowing marine sediments to accumulate. Even the presently high interior of the continent was once invaded by inland seas (Fig. 162), and thick marine sediments were deposited in the deep basins. When the seas departed and the land rose, erosion exposed many of these marine sediments along with their content of fossils.

Limestones are among the best rocks for finding fossils. This is due to their means of sedimentation, incorporating shells and skeletons of dead marine life that were buried and fused into solid rock. Most limestones are marine in origin with some deposits originating in lakes. Limestones constitute approximately 10 percent of all exposed sedimentary rocks. Shales are the most prominent sediments, followed by sandstones.

Many limestones form massive outcrops (Fig. 163), which are recognized by their typically light gray or light brown color. The application of a few drops of 10 percent hydrochloric acid solution (available at drugstores and rock shops) is a further test for limestone. The reaction of acid on calcium carbonate produces a strong effervescence on a fresh surface. This is also a good test for limy mudstones and sandstones because they are cemented with calcite.

Whole or partial fossils constitute most limestones, depending on whether they were deposited in quiet or agitated waters. Tiny spherical grains called oolites are characteristic of agitated water, whereas lithified layers of limy mud called micrite are characteristic of calm water. In quiet waters, undisturbed by waves and currents, whole organisms with hard body parts are buried in calcium carbonate sediments, which are later lithified into lime-

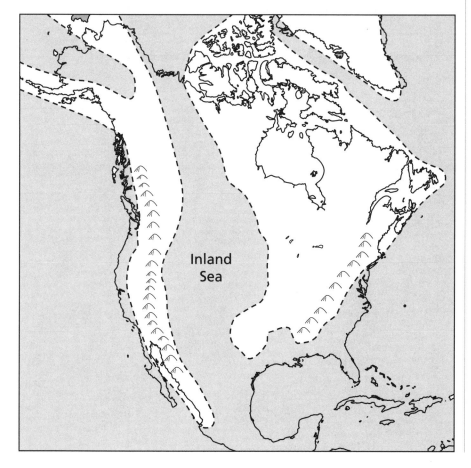

Inland
Sea

Figure 162 The paleo-geography of the Cretaceous period, showing a major inland sea.

Figure 163 *Limestone formation of the Bend Group in a ravine at the base of the Sierra Diablo escarpment, Culberson County, Texas.*

(Photo by P. B. King, courtesy USGS)

stone. In agitated waters near shore, shells and other hard body parts are fragmented by the back-and-forth motion of the waves and tides.

Most carbonate sediments were deposited in fairly shallow waters, generally less than 50 feet deep, and the majority of these were deposited in intertidal zones, where marine organisms are plentiful. Coral reefs, which form in shallow water, where sunlight can easily penetrate, contain abundant organic remains. Many ancient carbonate reefs are composed largely of carbonate mud, in which larger skeletal remains literally "float" in the mud.

Some carbonate rocks were deposited in deep seas; however, their fossil content is usually poor. The maximum depth at which carbonate rocks can form is controlled by the calcium carbonate compensation zone, which usually begins about two miles deep. Below this depth, the cold, high-pressure waters of the abyss, which contain the vast majority of free carbon dioxide, dissolve calcium carbonate that sinks to this level.

Carbonate rocks began as sandy or muddy calcium carbonate material. The sand-size particles are composed of broken up skeletal remains of invertebrates and shells of calcareous algae that rain down from above. Skeletal remains might have been broken by mechanical means, such as the pounding of the surf, or by the activity of living organisms. Further breakdown into dust-size particles produces a carbonate mud (sometimes called marl), which is the most common constituent of carbonate rocks.

Under certain conditions, carbonate mud might dissolve in seawater and redeposit elsewhere on the ocean floor as calcite ooze. If dissolved calcium carbonate is deposited around a small nucleus such as a skeletal fragment or a quartz grain, it grows by adding layers of concentric rings into a sand-size particle called an oolith.

As calcareous sediments accumulate in thick deposits on the ocean floor, deep burial of the lower strata produces high pressures, which lithify them into carbonate rock, consisting mostly of limestone or dolostone. If fine-grained calcareous sediments are not strongly lithified, they form deposits of soft, porous chalk. If the sediments consist mostly of fragments of skeletal remains, they are cemented into a limestone called coquina. Limestones typically develop a secondary crystalline texture, whereby crystals grow after formation of the original rock by solution and recrystallization. Fossils in the limestone also recrystallize and are often destroyed by this process.

Shales and mudstones commonly contain fossils and are the most abundant sedimentary rocks because they are the main weathering products of feldspars, the most abundant minerals. In addition, all rocks are eventually ground down to clay-size particles by abrasion. Because clay particles are small and sink slowly, they normally settle out in calm, deep waters far from shore. Compaction squeezes out the water between sediment grains, and the clay is lithified into shale, which is easy to recognize by its thin, fissile (easily split) layers. Organisms caught in the clay are compressed into thin carbonized remains or impressions.

The deep, calm bottom waters were probably stagnant and oxygen-poor. Periodic slumping from a high bank would result in a flow of mud into the deeper waters. Organisms living on or in the shallow muddy bottom would be caught in the slide and buried in the mud when the slide came to rest in deeper water (Fig. 164). Because scavengers cannot survive in these waters, the remains of the organisms were favored for preservation. As the mud gradually compacted and became hard rock, the buried carcasses were flattened into dark carbonized films. Fossilization in this manner can also preserve an animal's soft parts, which are less well preserved in limestone.

Sediments such as coarse marine sandstones generally are not as fossiliferous as limestones and shales, probably because a high influx of these sediments into the ocean tends to choke bottom-dwelling organisms. Therefore, areas with active sediment deposition are usually devoid of all but the most sparsely populated bottom dwellers. However, under certain conditions of catastrophic flooding or submarine slides, entire communities might be buried within a layer of sandstone.

Sandstones can provide excellent terrestrial fossils and imprints (Fig. 165) and faithfully record the passage of animals by their fossil footprints, especially those of large dinosaurs. These are best made with deep impressions in moist sand that are filled with sediment such as windblown sand and later

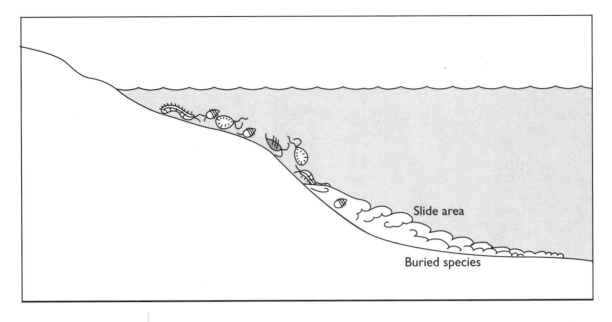

Figure 164 Undersea landslides played an important role in the fossilization of species.

buried and lithified into sandstone. Subsequent erosion exposes layers of sandstone, and the softer material that originally filled the depression weathers out, often exposing a clear set of footprints.

MINERAL-BEARING ROCKS

Most minerals that form crystals of collectible size are found in igneous rocks, some are found in sedimentary rocks, and a few are found in metamorphic

Figure 165 A large fossil leaf in a sandstone block found at the Coryell coal mines, Newcastle, Colorado.

(Photo by H. S. Gale, courtesy USGS)

222

rocks. Many metallic ore mines are excavated in igneous rocks; therefore, these rocks are a major source of the mineral wealth throughout the world. Most igneous rocks are aggregates of two or more minerals. For example, granite is composed almost entirely of quartz and feldspar with a minor constituent of other minerals. Granitic rocks formed deep inside the crust, and crystal growth was controlled by the cooling rate of the magma and the available space.

Large crystals probably formed late in the crystallization of a large magma body such as a batholith. They also might form in the presence of volatiles such as water and carbon dioxide, permitting large crystals to grow in a small magma body such as a dike (Fig. 166). As the magma body slowly cooled, possibly over a period of a million years or more, the crystals were able to grow directly out of the fluid melt or out of the volatile magmatic fluids that invaded the surrounding rocks.

If a granitic rock develops very large crystals, it is a pegmatite, which is the major source of single-mineral crystals throughout the world. Granite pegmatites are known for their crystals of enormous size. They also contain rare minerals with smaller crystals. The granite associated with pegmatites often consists of quartz rods in a matrix of feldspar combined in a curious interlocking angular pattern that resembles Egyptian hieroglyphic writing. For

Figure 166 *Vertical sheets of diorite split apart by intruding El Capitan granite dike on the floor of Tiltill Valley, Yosemite National Park, Tuolumne County, California.*

(Photo by F. E. Matthes, courtesy USGS)

this reason, it is called graphic granite and is an attractive rock found in large quantities where pegmatites are known to exist. Therefore, these rocks are a key to locating pegmatite deposits.

Pegmatite bodies are generally dikelike (tabular) and lens-shaped (lenticular) or consist of massive blobs within the parent rock. They range in length from a few inches to several hundred feet, and exceptional pegmatites have been traced for several miles. Pegmatites can be found in many localities where well-exposed granite outcrops exist. But they are especially prevalent in the eastern and the western mountainous regions of the United States.

Ordinary pegmatites are composed mainly of great masses of quartz and microcline (potassium) feldspar (Fig. 167), which are the dominant minerals in granite. Individual crystals range from a fraction of an inch to colossal size, weighing many tons. Quartz crystals weighing thousands of pounds have been found. Microcline crystals from Maine were reported to be up to 20 feet across. Some pegmatites might contain huge booklike segregations of mica over 10 feet wide. Also, large crystals of plagioclase (sodium) feldspar called albite might be present in a tabular or platy form.

Extraordinary pegmatites contain a gamut of minerals in large crystals of which beryl, tourmaline, topaz, and fluorite are the most common. Many minerals, such as apatite, monazite, and zircon, which normally form only microscopic crystals in granite, form especially large crystals in pegmatites. Pegmatites are mined extensively in many parts of the world for their large feldspar crystals, which are used in the ceramics industry. In the Karelia region of Russia, thousands of tons of feldspar were mined from several gigantic crystals, possibly the largest in the world.

Other industrial minerals, including radioactive ores, are mined from pegmatites. In addition, numerous gemstones and the so-called rare-earth elements, which are important in manufacture of superconductors, are associated with pegmatites. Some crystals in these rocks also reach enormous size. Examples are spodumene crystals found in the Black Hills of South Dakota that were over 40 feet long and beryl crystals in Albany, Maine, that measured as much as 27 feet long and six feet wide.

Volcanoes provide numerous samples of diverse welded tuffs, agglomerates, and ignimbrites (recrystallized ash flows) that are readily obtained by the collector. Ancient lava flows might contain clear, dark green, or black natural glass called obsidian. Some lava flows might contain cavities or vesicles that are filled with crystals called zeolites, meaning "boiling stones," because they formed when water boiled away as the basalt cooled. Trachytes, volcanic glass of basalt composition, often contain large well-shaped feldspar crystals that are aligned in the direction of the lava flow. Serpentine, so named because it often has the mottled green color of a serpent, is soft and easily polished, characteristics that allow it to be worked into a variety of ornaments and decorative objects.

Calcite, composed of calcium carbonate, is chief among the minerals precipitated from seawater. Under certain conditions, calcite can form large, lustrous crystals. One of the largest calcite crystals was found in Sterling Bush, New York; it measured 43 inches long and weighed about half a ton. Calcite can grow in cavities to form long, tapering crystals that resemble dogs' teeth, called dogtooth spar. Mineral springs produce layered deposits of calcium carbonate called travertine, which, because of numerous cavities, resembles Swiss cheese. Crystals of a particularly clear form of calcite have optical qualities and have been used in petrographic, or rock, microscopes and aircraft bomb sights.

In limestone caves, beautiful stalactites hang from the ceiling and stalagmites cling to the floor (Fig. 168). They resemble long "icicles" of calcite that grow by the precipitation of acidic groundwater seeping through the rock. Stalactites are also called dripstones because they formed by dripping water containing dissolved calcium carbonate. The process is extremely slow, and a period of hundreds of years is required for stalactites and stalagmites to grow a single inch. Sometimes the two formations meet to form columns.

Many caves also contain exquisite twisting fingers of calcite or aragonite called helictites. They form as stalactites do except that water drips through them too slowly for drops to form, creating contorted, branching cave deposits. The moisture reaches the tip of the helictite and evaporates, so that crystals do

Figure 168 *Stalactites in Carlsbad Caverns, Eddy County, New Mexico.*

(Photo by W. T. Lee, courtesy USGS)

Figure 169 *Gypsum crystals in carbonaceous shale bed, Powder River County, Montana.*

(Photo by C. E. Dobbin, courtesy USGS)

not grow straight down but in curls and spirals. Aragonite, which is similar to calcite but has a different crystal structure, forms rough, needlelike helictites.

Some cave deposits include cave coral, created when water reaches the cave through a network of channels too small to allow drops to form. Instead, moisture is squeezed out onto the cave wall, where it evaporates, leaving bumpy deposits that grow into various shapes that resemble popcorn, grapes, potatoes, and cauliflower. A drapery is formed when water droplets flow down a sloping cave ceiling, leaving trails of calcite that build up layer by layer. Sheets of calcite, called flowstone, are deposited when wide streams of water flow down cave walls, often constructing massive terraces. In underwater caves, such as those on the Yucatan Peninsula in Mexico, the limestone formations also include delicate, hollow stalactites called soda straws that took millions of years to create but are destroyed in mere moments by careless divers.

Thick beds of gypsum, composed of hydrous calcium sulfate, are among the most common sedimentary rocks (Fig. 169). They formed in evaporite deposits, which occur when a pinched off portion of the ocean or an inland sea evaporates. Oklahoma, like many parts of the interior of North America that were invaded by a Mesozoic sea, is well known for its gypsum beds. Evaporite deposits contain large amounts of halite as well and are mined extensively throughout the world for this mineral. Gypsum is mainly used for the manufacture of plaster of paris and wallboard. It often develops into large, chevron-shaped double crystals called swallowtail twins. When gypsum is heated or compressed, water is driven off, and the mineral becomes anhydrite, which can revert to gypsum in the presence of water. If massive gypsum is compacted, it forms alabaster, which is often carved into vases and ornaments.

Sedimentary rocks frequently contain concretions, which are mineralized nodules. They occur in a variety of shapes, sizes, and colors. If the nodules are hollow and are lined with agate or various other crystals, they form geodes. When cut open, geodes reveal small, bright crystals or bands of colorful agate (Fig. 170). Geodes also can occur when crystals composed of opal or chalcedony fill large cavities in basalt; for this reason, they are called "thunder eggs." Geodes make beautiful specimens when cut and polished.

If clay derived from the weathering of feldspar is buried deep inside the Earth, the extreme temperatures and pressures can alter it into muscovite, garnet, and other metamorphic minerals. Metamorphism also takes place in contact with hot magma bodies that invade the crust, and for this reason the process is called contact metamorphism. During the metamorphic process, minerals in either sedimentary or igneous rocks might undergo recrystallization, which forms larger crystals. Chemical changes might take place, in which new minerals are formed in the place of old ones. Among the minerals found in contact metamorphic zones that might be of interest to collectors are garnet, epidote, and diopside.

GEOLOGIC MAPS

In order to locate areas where fossils and minerals are found, a basic understanding of geologic maps is useful. All fossils and minerals occur in specific rock types and often become important components of the rocks themselves. Some geologic forms and structures are associated with a particular rock type and can often be recognized some distance away. Therefore, the size, shape, and composition of landforms depend on the nature of the rocks that constitute them. Rocks that form mountains or underlie valleys become visible when

Figure 170 Types of silica-lined geodes from green glass. At left is a geode with opal and chalcedony; center is a Dugway geode weathered from green glass and rounded by rolling on beaches; right is the interior of a Dugway geode showing banded agate and the hollow center lined with drusy quartz crystals.

(Photo by M. H. Staatz,

228

Figure 171 *The snow-covered peaks of the Rocky Mountains near Telluride, Colorado.*

(Photo by W. Cross, courtesy USGS)

they are pushed or folded upward or when they jut through the ground and become outcrops.

Landforms, including cliffs, hills, canyons, valleys, plateaus, and basins, are the basic features of the Earth. They often express the type of rock that they comprise, which in turn determines whether they contain collectible fossils or minerals. All landforms are the result of a combination of processes that continually build up the land surface and tear it down. Knowledge of landforms and structures is therefore fundamentally important to exploration for fossils and minerals and aids in interpreting the landscape and the geologic history of an area.

The Earth's surface has been fashioned by a variety of geologic forces, offering a wide assortment of structures. It is constantly changing, sculptured by such formidable geologic processes as plate tectonics, uplift and erosion, powerful earth-moving processes, and catastrophic collapse. Geologic structures resulting from the interaction of movable tectonic plates, igneous activity, and ground motions constitute many unusual land features. Spectacular landscapes are carved by unceasing weathering and erosional agents that cut down the tallest mountains and gouge out the deepest ravines. Rivers transport the sediments to the sea, where they lithify into solid rock and are exposed on dry land. The presence of these geologic phenomena is a tribute to the powerful tectonic forces responsible for creating a large variety of landforms.

Mountain ranges, created by the forces of uplift and erosion, produce some of the most rugged terrain (Fig. 171). They are towering topographical features, involving massive deformation of the rocks in their cores. Mountains contain complex internal structures formed by folding, faulting, volcanism, igneous intrusion, and metamorphism. Continental collisions built folded mountain belts equally imposing. The surface is also riddled with faults that stack blocks of crust into mountainous terrain. These regions offer some of the best exposures for locating fossils and minerals.

Volcanoes also built chains of mountains and are further expressions of the tireless tectonic activities that constantly reshape the surface of the Earth. More than three-quarters of the Earth's surface above and below the sea is of volcanic origin. Volcanoes are the most spectacular of all Earth processes. They add material to the crust, thereby building new landscapes. Thus, volcanic eruptions play a major role in the growth of the continents. Collapsing volcanoes leave huge gaping calderas that, when filled with water, form some of the world's deepest lakes.

The Earth is constantly evolving, with complex activities such as the flow of water and the motion of waves. Running water is responsible for sculpting the landscape more than any other natural process. Weathering, down-slope movement, and river flow work together to reshape the continents. A veneer of sediments covering the surface is fashioned by erosion into plateaus and mesas, canyons and valleys. Rivers erode valleys and provide a system of drainage delicately balanced with climatic, topographic, and lithologic characteristics. Even in the most arid regions, the principal topographic features result from excavation by stream erosion. Blowing wind forms desert landscapes and powers roving sand dunes. Gigantic dust storms and sandstorms prevalent in desert regions also play a major role in shaping the arid landscape.

Sediments eroded from the highlands and carried by rivers to the sea carve out new landforms. Shifting sediments on the land surface along with the accumulation of deposits on the ocean floor continuously remold the planet. Rivers deliver to the sea a heavy load of sediments washed off the continents, continually building up the coastal regions. Seacoasts vary dramatically in topographic characteristics, climate, and vegetation. They are places where continental and oceanic processes converge to produce a landscape that is invariably changing on a rapid scale.

These landforms are viewed in plan form by a topographic map, which is simply a drawing indicating surface features in detail. It is a means of viewing the Earth's surface in three dimensions by drawing landforms using contour lines of equal elevations (Fig. 172). Topographic maps show more detail of natural and human-made features than ordinary two-dimensional maps by depicting relative positions along with elevations. The maps are useful to geologists, civil engineers, and many other users of maps who require that elevations of landforms such as hills and valleys be indicated. They are especially

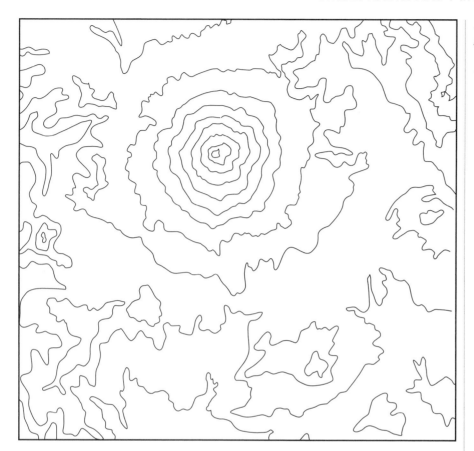

Figure 172 *Rough topographic map of Mount Saint Helens prior to the May 1980 eruption.*

useful in determining whether geologic structures have changed significantly over time, such as the erosion of beaches and sea cliffs. The maps not only reveal the present form of the surface but also might give useful information about past events that have shaped the region.

Topographic maps are primarily used for military, scientific, and commercial purposes. The U.S. Geological Survey (USGS) has mapped most of the United States and made the resulting maps available to the public. They are small-scale topographic sheets or quadrangles usually with scales of one inch equal one mile or less. Symbols are used on the topographic sheets to represent natural features as well as man-made structures. The maps are also printed in various colors such as brown for contour lines, green for vegetation, blue for bodies of water, red for highways and boundary lines, and black for artificial works.

Earlier topographic maps were made largely from ground observations. Today, these maps are made entirely from aerial photographs and then are field-checked for accuracy. The maps are drawn by using two aerial photographs, called stereo pairs, taken at different points along the flight line, thereby viewing objects on the ground at two different angles, producing a three-dimensional image. Topographic maps are made from these images by selecting points of equal elevation. Stereo photographs also can be made from satellite images viewed from different angles.

The maps are drawn with contour lines of equal elevations that are often shaded with various colors to emphasize relief. A contour interval of 10 or 20 feet is typically used on most topographic maps. In maps of mountainous terrain, a contour interval of 50 or 100 feet might be used to prevent crowding the contour lines on steep slopes. In extremely flat areas, the contour interval might only be a few feet. Generally, each fifth contour line is printed in a slightly darker color for emphasis. The maps are made with an accuracy that requires that 90 percent of all well-defined features be within a fraction of an inch of their true location on the ground and that the elevations of 90 percent of the features be correct within one-half of the contour interval.

Widely and evenly spaced contour lines indicate land that is regular and only slightly sloping, whereas irregular contour lines spaced close together show land that is rough and steeply sloping. Features such as hills, valleys, canyons, ridges, and cliffs are easily recognized from their distinctive contour lines. Hills are represented by a series of concentric closed curves, whereas valleys and canyons are represented by a series of U- or V-shaped contours that bend upstream. Ridges and cliffs are represented by long parallel contours spaced fairly closely together. Features such as basins or craters appear as a series of concentric closed curves like those that represent hills, but their contour lines are drawn with hash marks pointing inward to prevent confusion.

One of the most practical uses of fossils was in delineating rock formations, which in turn were utilized for geologic mapping. A geologic map is important for displaying various rock types on the surface. It presents in plan view the geologic history of an area where particular rock exposures are found. The purpose of geologic maps is to display the distribution of rocks on the Earth's surface. The maps also indicate the relative ages of these formations and profile their position underground. Often, much of the information is compiled from just a few available exposures, which must be extrapolated over a large area. The first geologic maps were made by geologists in Britain, where one of the most practical uses was in the exploration of coal. In the western United States, pioneer geologists made extensive geologic maps, often sketching rock formations on horseback while riding through the region.

Modern geologic maps incorporate field observations and laboratory measurements, which are limited by rock exposures, accessibility, and personnel. Regional geologic maps present rock composition, structure, and geologic age, which are essential for constructing the geologic history of an area. Aided by geophysical data used for defining subsurface structures, these reconstructions are important because formations of rock units and geologic structures influence the deposition of much of the mineral wealth of the world.

Airplane and satellite remote sensing methods are primarily used for augmenting conventional techniques of compiling and interpreting geologic maps of large regions. By using remote sensing techniques, certain structural and lithologic information is obtained much more efficiently than can be achieved on the ground. In well-exposed areas, geologic maps can be made from aircraft and satellite imagery, even when only limited field data are available because many of the major structural and lithologic units are well displayed on the imagery. The information provides a means of generating large-scale geologic maps of inaccessible areas as well.

Long, linear trends in the Earth's surface, called lineaments, are among the most obvious as well as most useful features in the imagery. Lineaments represent zones of weakness in the Earth's crust often as a result of faulting. Lineaments and texture along with dip and strike of the strata, which are the degree and direction of formation slope, further aid in geologic mapping. Even with subdued eroded topographic features, major geologic structures can be discerned.

Other features frequently observed in the images include circular structures created by domes, folds, and intrusions of igneous bodies into the crust, along with lineaments, joints, fracture patterns, and other erosional features. Structural features such as folds, faults, dips, and strikes of particular rock formations, along with lineaments, landform features, drainage patterns, and other anomalies, might suggest areas where minerals and petroleum can be found.

Stream drainage patterns provide important information about the geologic structure and characteristics of an area. Streams and the valleys they cut join into networks that display various types of drainage patterns, depending on the terrain. If the terrain has a uniform composition and does not determine the direction of valley growth, the drainage pattern is dendritic (Fig. 173), resembling the branches of a tree. Granitic and horizontally bedded sedimentary rocks generally yield this type of stream pattern.

A trellis drainage pattern displays rectangular shapes that reflect differences in the bedrock's resistance to erosion. The major tributaries of the master stream follow parallel zones of least resistance over underlying folded rocks. Rectangular drainage patterns also occur when fractures crisscross the bedrock, forming zones of weakness that are particularly susceptible to erosion. If streams radiate outward in all directions from a topographical high such as a volcano or dome, they produce a radial stream drainage pattern.

One of the best clues to the identity of a geologic feature is its surface relief. Stream drainage patterns are influenced by topographic relief and rock type. In areas of exposed bedrock, drainage patterns depend on the lithologic features of the underlying rocks, the attitude of rock units, and the arrangement and spacing of planes of weakness encountered by runoff. In addition, the color and texture of the structure impart information about the rock formations that comprise it.

233

Figure 173 *Dendritic drainage pattern in an area underlain by Gila conglomerate, Gila County, Arizona.*

(Photo by N. P. Peterson, courtesy USGS)

Surface expressions such as domes, anticlines, synclines, and folds bear clues about the subsurface structure. Various types of drainage patterns imply variations in the surface lithologic characteristics or rock type. The drainage pattern density is another indicator of lithologic features. Variations in the drainage density also correspond to changes in the bedrock and coarseness of the alluvium. Any abrupt changes in the drainage patterns are particularly important because they signify the boundary between two rock types, which might indicate mineral emplacement.

COLLECTING FOSSILS AND MINERALS

Geology is one of the few scientific fields in which the amateur can still contribute substantially to its progress. Important discoveries are made all the time, not only by professional scientists but also by serious amateur geologists who seek the thrill of discovery. However, certain precautions must be observed when collecting fossils and minerals. When exploring for specimens on public lands such as national parks, national monuments, national forests, wilderness areas, and state lands, ask a park ranger whether any restrictions apply. Also, if collecting on private land, obtain permission from the landowner or mine owner. In addition, some states might require a permit to collect fossils.

Of particular importance is the avoidance of sites where scientific work is being conducted. Major paleontological digs give amateur fossil hunters an opportunity to work alongside professionals in excavating dinosaur bones or fossils of other species. Organizations offer field trips for those who are interested in paleontology but have no formal training. Some museums and parks even allow visitors to take part in paleontological digs. Also, in remote parts of the world, volunteer workers contribute substantially to scientific research.

The search begins by visiting the local library. Many libraries have materials such as geologic road guides, other detailed guidebooks, and hobby books to acquaint the novice with the practicalities of fossil and rock collecting. The division of mines in most states, the U.S. Bureau of Mines, the U.S. Geological Survey, and the Bureau of Land Management might provide additional information. Many colleges and universities have geology departments that can assist the amateur collector. Natural history museums contain a wealth of information. Word of mouth from other collectors is also a good source of information. Often, rock shops are excellent places for purchasing books and equipment and obtaining information on the best sites for collecting.

Little equipment is required for successful rock collecting, and it might be found lying around the house or bought from a hardware store. A map, compass, notebook, and perhaps a field guide might come in handy for locating a collecting site. A geologist's pick or a bricklayer's hammer and a sizable chisel can be used for breaking and splitting rocks. A trowel or a large knife can be used for splitting shale. A sledge hammer, shovel, and pry bar might come in handy for excavating rock. A sieve or strainer can be used for sifting loose sediments. A hand lens or a magnifying glass might be required for observing minute specimens. A paintbrush, toothbrush, or whisk broom can be used for sweeping clean the work space. A knapsack and various types of packing containers and newspaper for wrapping specimens can be used to transport findings home safely. Labeling and numbering specimens to site location are also important.

Specimens should be taken home to be cleaned to save time and prevent breakage in the field. Fossils can be cleaned by washing with detergent and water. The enclosing rock can be removed with a knife, chisel, or saw. Sharp, pointed implements can be used to clean small specimens. Specimens encased in limestone can be freed by dissolving the rock in a weak solution of hydrochloric or sulfuric acid.

Figure 174 Limestone
quarry in Franklin
County, Alabama.

(Photo by E. F. Burchard,
courtesy USGS)

Because limestone is used in the manufacture of Portland cement for
building and road construction, numerous limestone quarries dot the landscape across the country (Fig. 174). In addition, building materials and gravel
for roads are made from limestone, and limestone gravel pits are a common
sight in many parts of the United States, though caution should be used in the
vicinity of any mine or pit. These quarries are often ideal sites for collecting
fossils because many layers of limestone might be penetrated. In addition, the
rocks have been conveniently broken up, exposing fresh surfaces, on which
fossils, especially abundant shells, are clearly seen.

Coal pits are also numerous and are excellent locations for collecting fossils of plants and animals that were buried in the great ancient coal swamps.
Extensive forests and swamps grew successively on top of one another and continued to add to thick deposits of peat, which were buried under layers of
sediment and compressed into lignite, bituminous, and anthracite coal. Most
specimens are carbonized imprints of plant leaves and stems between easily spilt
layers of coal or associated shale and mudstone. Some coal seams or inter-
bedded shales contain minerals of collectible quality. The semiprecious gem jet
is a particularly tough form of coal that can be polished to a high sheen.

Most ores are associated with other minerals, some of which have good
crystals. Abandoned ore mines, especially those in the West (Fig. 175), are some
of the best locations for collecting rocks and minerals, and, indeed, some of

the best mineral specimens found in museums and in other collections were taken from mines. As mentioned, the largest crystals have been discovered in mines that worked granite pegmatites. Open-pit mines provide large exposures of rock that can be explored for minerals. The waste dumps of abandoned underground mines are another excellent source of minerals. A mine shaft might penetrate a particularly well-crystallized zone on its way to an ore body, and these rocks, which are of no use to miners, end up in the waste dumps. They are much safer than the mines themselves, which are extremely dangerous and prone to caving and should be avoided. Adjacent outcrops might also be mineralized and could yield valuable specimens.

Road cuts are perhaps the most accessible locations for exploring for fossils and minerals because the collector can simply drive to them. Again, however, caution must be taken. When a road is constructed through hilly or mountainous terrain, often huge quantities of rock have to be blasted out, providing excellent exposures. Many of the geologic features across the United States have been mapped by using road cuts, railway cuts, and tunnels, which are sometimes the only good rock exposures. If a road is cut through limestones or shales, fossils might be present. If a road is cut through igneous or metamorphic rocks, minerals, including some with sizable crystals, can be found. Table 11 (page 240) indicates locations where relatively abundant fossils of several animals and plants can be found in the United States.

Of lesser accessibility but with equally good, if not better, rock exposures are stream channels, especially those that cut deep through hard rock. The Grand Canyon (Figs. 176, 177) provides perhaps the best example; it is carved

Figure 175 *Lille and Vindicator mines in Cripple Creek Mining District, Colorado, in 1903.*

(Photo courtesy USGS)

Figure 176 *A view of the Grand Canyon from Mohave Point.*

(Photo by George A. Grant, courtesy National Park Service)

Figure 177 *A cutaway view of rock bed formations that comprise the Supai Group of the Grand Canyon, Arizona.*

(Courtesy USGS Professional Paper 1173)

through rocks that are hundreds of millions of years old and over a mile thick. The Grand Canyon lies in the southwest end of the Colorado Plateau, a generally mountain-free expanse that stretches from Arizona north into Utah and west into Colorado and New Mexico. The region was uplifted during the rising of the Rocky Mountains.

Between 10 and 20 million years ago, the Colorado River began eroding layers of sediment, exposing the raw basement rock below, and its present course is less than 6 million years old. Well-exposed sediment layers tell a near-complete story about the geologic history of the Earth. On the bottom of the canyon lies the ancient original basement rock, upon which sediments were slowly deposited layer by layer. Exposed on the wall of the canyon is the Great Unconformity, one of America's most prominent and famous geologic features, where relatively young rocks lie atop rocks that are much older.

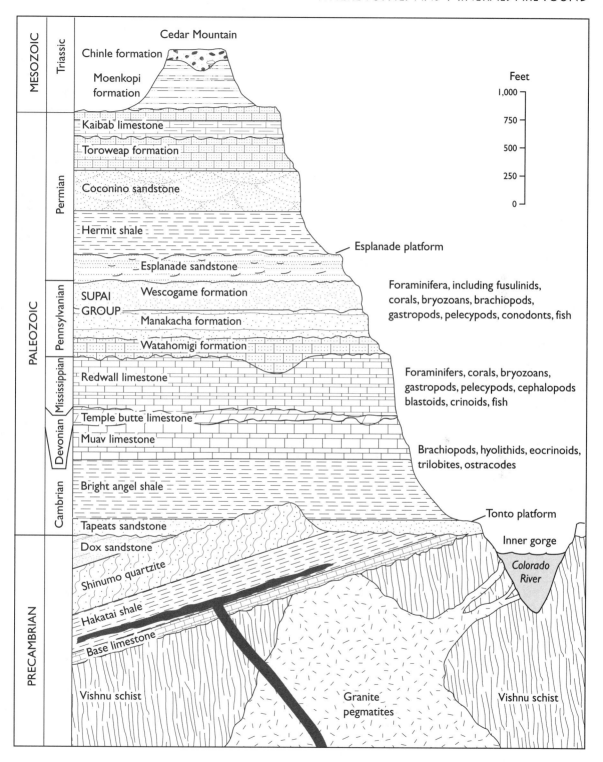

MESOZOIC	Triassic	Cedar Mountain
		Chinle formation
		Moenkopi formation

Feet
1,000
750
500
250
0

Kaibab limestone
Toroweap formation
Coconino sandstone
Hermit shale

Esplanade platform

Esplanade sandstone

SUPAI GROUP
Wescogame formation
Manakacha formation
Watahomigi formation

Foraminifera, including fusulinids, corals, bryozoans, brachiopods, gastropods, pelecypods, conodonts, fish

Redwall limestone

Temple butte limestone
Muav limestone

Foraminifers, corals, bryozoans, gastropods, pelecypods, cephalopods blastoids, crinoids, fish

Bright angel shale

Brachiopods, hyolithids, eocrinoids, trilobites, ostracodes

Tapeats sandstone

Tonto platform

Dox sandstone

Inner gorge

Shinumo quartzite

Colorado River

Hakatai shale

Base limestone

Vishnu schist

Granite pegmatites

Vishnu schist

PALEOZOIC — Permian — Pennsylvanian — Mississippian — Devonian — Cambrian

PRECAMBRIAN

239

The easiest fossil collecting is in areas where specimens have been weathered out of the rock and lie in loose sediment or rocky debris, called scree or float, at the base of an exposure. Often the limestone encasing a fossil erodes more easily, leaving whole specimens scattered on the ground. Broken-up rock in abandoned limestone quarries might also provide good hand samples that contain abundant fossils. A rock containing a desirable fossil might have to be chiseled out of a large boulder. However, great care should be exercised when attempting to remove a fossil from hard rock so that the specimen does not crack or crumble and thus destroy what nature has preserved for millions of years.

TABLE 11   FOSSILS BY STATE

Ammonites

Alabama
Alaska
Arkansas
California
Illinois
Kansas
Montana
Nebraska
Nevada
New Jersey
Oklahoma
Oregon
South Dakota
Texas

Brachiopods

Alabama
Arkansas
California
Colorado
Georgia
Idaho
Iowa
Maine
Massachusetts
Missouri
Montana
Nebraska
New Hampshire
New Jersey
New York
Ohio
Oregon
Pennsylvania
Tennessee
Virginia
West Virginia

Bryozoans

Alabama
Georgia
Nebraska
Pennsylvania
Texas
West Virginia

Cephalopods

Colorado
Delaware
Georgia
Michigan
Mississippi
Missouri
West Virginia

Coral

Alabama
California
Georgia
Illinois

Michigan
Mississippi
New York
Pennsylvania
Texas
Virginia
West Virginia

Crinoids

Colorado
Iowa
Missouri
Montana
New York
Oklahoma
Oregon
Pennsylvania
Texas
Virginia
West Virginia
Wyoming

Dinosaurs

Arizona
Colorado
Connecticut
Massachusetts
Montana
New Jersey
New Mexico

Oklahoma
Tennessee
Texas
Utah
Wyoming

Fish

Alaska
Arkansas
California
Connecticut
Kansas
Massachusetts
New Jersey
New York
Ohio
Texas
Virginia

Gastropods

Alabama
California
Colorado
Delaware
Georgia
Mississippi
Missouri
New Hampshire
Oregon
Pennsylvania

Tennessee
Texas
West Virginia

Graptolites

Alabama
Arkansas
Idaho
Maine
New York
Pennsylvania
Virginia
Washington

Insects

Alaska
California
Colorado
Illinois
Kansas
Montana
New Jersey
Oklahoma

Mammoths

Alaska
Florida
Nebraska
New York
South Dakota

Mastodons

Florida
Nevada
New Jersey
New York
Ohio

Mollusks

Alabama
Alaska
Massachusetts
Mississippi
New Jersey
New York
Oregon

Pelecypods

Alabama
Colorado
Delaware
Florida
Kansas
Louisiana
Mississippi
Missouri
Montana
Nebraska
Ohio
Pennsylvania
Texas

Virginia
Washington

Petrified wood

Arizona
California
Colorado
Idaho
Kansas
Montana
Nebraska
Nevada

Oklahoma
Oregon
South Dakota
Texas
Washington
Wyoming

Plants

Alabama
Arkansas
California
Colorado
Georgia

Illinois
Indiana
Iowa
Kansas
Kentucky
Massachusetts
Mississippi
Montana
New Mexico
North Carolina
North Dakota
Ohio
Oregon
Pennsylvania

Texas
Virginia
Washington
West Virginia

Shark's teeth

Alabama
Arkansas
California
Colorado
Florida
Maryland

Mississippi
New York
Oklahoma
Oregon
South Carolina

Trilobites

Alabama
California
Idaho
Illinois
Indiana
Iowa

Kansas
Massachusetts
Minnesota
Missouri
Nebraska
New Hampshire
New York
Ohio
Pennsylvania
Tennessee
Vermont
Virginia
Washington
West Virginia

GLOSSARY

aa lava (AH-ah)   Hawaiian name for blocky basalt lava flow

abyss   the deep ocean, generally over a mile in depth

acanthostega (ah-KAN-the-stay-ga)   an extinct primitive Paleozoic amphibian

agate   a fine-grained variety of quartz composed of varicolored bands of chalcedony, usually occurring within rock cavities

age   a geologic time interval that is shorter than an epoch

amber   fossil tree resin that has achieved a stable state after ground burial, through chemical change and the loss of volatile constituents

ammonite (AM-on-ite)   a Mesozoic cephalopod with flat, spiral shells

amphibian   a cold-blooded, four-footed, vertebrate, belonging to a class midway in evolutionary development of fish and reptiles

amphibole (AM-feh-bowl)   a group of complex silica minerals containing calcium, magnesium, and iron

andesite   a volcanic rock intermediate between basalt and rhyolite

angiosperm (AN-jee-eh-sperm)   flowering plant that reproduces sexually with seeds

anhydrite   a calcium sulfate mineral with similar composition to gypsum without water

annelid (A-nil-ed)   wormlike invertebrate characterized by segmented body with a distinct head and appendage

aragonite   a calcium carbonate mineral with composition similar to that of calcite but denser and harder with a different crystalline structure

242

archaeocyathan (AR-key-ah-sy-a-than) an ancient Precambrian organism resembling sponges and coral; they built early limestone reefs.

Archaeopteryx (AR-key-op-teh-riks) primitive Jurassic crow-size bird characterized by teeth and bony tail

archea (AR-key-ah) primitive bacterialike organism living in high-temperature environments

Archean (AR-key-an) major eon of the Precambrian from 4.0 to 2.5 billion years ago

arkose a coarse-grained sandstone with abundant feldspar

arthropod (AR-threh-pod) the largest group of invertebrates, including crustaceans and insects, characterized by segmented bodies, jointed appendages, and exoskeletons

augite (AH-jite) the most common pyroxene comprising black or dark green crystals in igneous rocks

Baltica (BAL-tik-ah) an ancient Paleozoic continent of Europe

basalt a dark volcanic rock rich in iron and magnesium and usually quite fluid in the molten state

basement the surface beneath which sedimentary rocks are not found; the igneous, metamorphic, granitized, or highly deformed rock underlying sedimentary rocks

batholith the largest of intrusive igneous bodies, more than 40 square miles on its uppermost surface

belemnite (BEL-em-nite) an extinct Mesozoic cephalopod with bullet-shaped internal shell

bicarbonate an ion created by the action of carbonic acid on surface rocks; marine organisms use the bicarbonate along with calcium to build supporting structures composed of calcium carbonate.

biotite a black or dark green mica in crystalline rocks

bivalve a mollusk with a shell comprising two hinged valves, including oysters, muscles, and clams

blastoid an extinct Paleozoic echinoderm similar to a crinoid with a body resembling a rosebud

brachiopod (BRAY-key-eh-pod) marine, shallow-water invertebrate with bivalve shells similar to mollusks and plentiful in the Paleozoic

breccia (BRE-cheh) a coarse-grained clastic rock with sharp fragments often embedded in clay

bryophyte (BRY-eh-fite) nonflowering plants comprising mosses, liverworts, and hornworts

bryozoan (BRY-eh-zoe-an) a marine invertebrate that grows in colonies and is characterized by a branching or fanlike structure

calcite a mineral composed of calcium carbonate

Cambrian explosion a rapid radiation of species that resulted from a large adaptive space, including numerous habitats and mild climate

carbonaceous (KAR-beh-NAY-shes) a substance containing carbon, namely, sedimentary rocks such as limestone and certain types of meteorites

carbonate a mineral containing calcium carbonate such as limestone and dolomite

Cenozoic (SIN-eh-zoe-ik) an era of geologic time comprising the last 65 million years

cephalopod (SE-feh-lah-pod) marine mollusks including squid, cuttlefish, and octopus that travel by expelling jets of water

chalcedony (kal-SEH-din-ee) a variety of quartz with a compact fibrous structure and waxy luster, often found as a deposit, lining, or filling in rocks

chalk a soft form of limestone composed chiefly of calcite shells of microorganisms

chert an extremely hard cryptocrystalline quartz rock resembling flint

chondrule (KON-drule) rounded granules of olivine and pyroxene found in stony meteorites called chondrites

cladistics a taxonomic system using evolutionary relationships to classify organisms

class in systematics, the category of plants and animals below a phylum comprising several orders

clastic a rock comprising fragments of other rocks

cleavage the tendency of a mineral to break along a plane due to a direction of weakness in the crystal

coal a fossil fuel deposit originating from metamorphosed plant material

coelacanth (SEE-leh-kanth) a lobe-finned fish originating in the Paleozoic and presently living in deep seas

coelenterate (si-LEN-teh-rate) multicellular marine organisms, including jellyfish and corals

conglomerate a sedimentary rock composed of welded fine-grained and coarse-grained rock fragments

conodont a Paleozoic toothlike fossil probably from an extinct marine vertebrate

coral a large group of shallow-water, bottom-dwelling marine invertebrates comprising reef-building colonies common in warm waters

coprolite fossilized excrement generally black or brown used to determine eating habits of animals

coquina (koh-KEY-nah) a limestone composed mostly of broken pieces of marine fossils

craton the ancient, stable interior region of a continent, usually composed of Precambrian rocks

crinoid (KRY-noid) an echinoderm with a flowerlike body atop a long stalk of calcite disks

crossopterygian (CROS-op-tary-gee-an) extinct Paleozoic fish thought to have given rise to terrestrial vertebrates

crustacean (KRES-tay-shen) an arthropod characterized by two pairs of antennalike appendages forward of the mouth and three pairs behind it, including shrimp, crabs, and lobsters

crystal a solid body with a regularly repeating arrangement of atomic constituents; the external expression might be bounded by natural planar surfaces called faces

diatom microplants whose fossil shells form siliceous sediments called diatomaceous earth

dike a tabular intrusive igneous body that cuts across the layering or structural fabric of the host rock

dinoflagellate (DIE-no-FLA-jeh-late) planktonic single-celled organisms important in marine food chains

diopside a pyroxene found in contact metamorphic rocks

dolomite a mineral formed when calcium in limestone is replaced by magnesium

echinoderm (i-KY-neh-derm) marine invertebrates, including starfish, sea urchins, and sea cucumbers

echinoid (i-KY-noid) a group of echinoderms including sea urchins and sand dollars

Ediacaran a group of unique extinct late Precambrian organisms

eon the longest unit of geologic time, roughly about a billion years or more in duration

epidote a calcium aluminum silicate, yellow-green mineral sometimes cut as a gem

epoch a geologic time unit shorter than a period and longer than an age

era a unit of geologic time below an eon, consisting of several periods

erathem a stratigraphic system consisting of rocks formed during an era

erosion the wearing away of surface materials by natural agents such as wind and water

esker a long narrow ridge of sand and gravel from glacial outwash stream

eukaryote (yu-KAR-ee-ote) a highly developed organism with a nucleus that divides genetic material in a systematic manner

eurypterid (yu-RIP-teh-rid) a large Paleozoic arthropod related to the horseshoe crab

evaporite the deposition of salt, anhydrite, and gypsum from evaporation in an enclosed isolated basin of seawater

evolution the tendency of physical and biologic factors to change with time

exoskeleton the hard outer protective covering of invertebrates including cuticles and shells

extinction the loss of large numbers of species over a short duration, sometimes marking the boundaries of geologic periods

extrusive   an igneous volcanic rock ejected onto the Earth's surface

family   in systematics, the category of plants and animals below order, comprising several genera

feldspar   a group of rock-forming minerals comprising about 60 percent of the Earth's crust and essential component of igneous, metamorphic, and sedimentary rock

fissile   the ability of a rock to split into thin sheets

fluvial   stream-deposited sediment

foliation   a planar arrangement of the textural or structural features in rocks

foraminifer (FOR–eh–MI–neh-fer)   a calcium carbonate–secreting organism that lives in the surface waters of the oceans; after their death, their shells form the primary constituent of limestone and sediments deposited on the seafloor

formation   a combination of distinct rock units that can be traced over a distance

fossil   any remains, impression, or trace in rock of a plant or animal of a previous geologic age

fulgurite (FUL-je-rite)   a tubular vitrified crust created by fusion of sand by lightning, most common on mountaintops

fumarole   a vent through which steam or other hot gas escapes from underground such as a geyser

fusulinid (FEW-zeh-LIE-nid)   a group of extinct foraminifers resembling a grain of wheat

gastrolith (GAS-tra-lith)   a stone ingested by an animal that is used to grind food

gastropod (GAS-tra-pod)   a large class of mollusks, including slugs and snails, characterized by a body protected by a single shell that is often coiled

genus   in systematics, the category of plants and animals below family, comprising several species

geode   a hollow, globular, nearly spherical mineral body in limestone and lavas

geologic column   the total succession of geologic units in a region

glauconite (GLO-keh-nite)   a green hydrous potassium iron silicate mineral found in greensands

glossopteris (GLOS-opt-ter-is)   a late Paleozoic plant that existed on the southern continents but not on the northern continents, thereby confirming the existence of Gondwana

gneiss (nise)   a foliated or banded metamorphic rock with composition similar to that of granite

Gondwana (GONE-wan-ah)   a southern supercontinent of Paleozoic time, which comprised Africa, South America, India, Australia, and Antarctica and broke up into the present continents during the Mesozoic era

granite   a coarse-grained, silica-rich igneous rock consisting primarily of quartz and feldspars

graptolite (GRAP-teh-lite)   extinct Paleozoic planktonic animals resembling tiny stems

graywacke (GRAY-wack-ee)   a coarse dark-gray sandstone

greenstone   a green weakly metamorphic igneous rock

gypsum   a common, widely distributed mineral frequently associated with halite or rock salt

Hadean eon   a term applied to the first half billion years of Earth history

hallucigenia (HA-loose-ah-gen-ia)   an unusual animal of the early Cambrian with stiltlike legs and multiple mouths along the back

helictite   a calcite branching cave deposit formed by water seeping through bedrock

hematite (HE-meh-tite)   a red iron-oxide ore

hexacoral   coral with six-sided skeletal walls

hornblende   a dark green or black amphibole important in metamorphic rocks

hornfels   a fine-grained siliceous rock produced by the metamorphism of slate

hydrocarbon   a molecule consisting of carbon chains with attached hydrogen atoms

hydrothermal   relating to the movement of hot water through the crust

Iapetus Sea (EYE-ap-i-tus)   a former sea that occupied a similar area as the present Atlantic Ocean prior to the assemblage of Pangaea

ichthyosaur (IK-the-eh-sore)   extinct Mesozoic aquatic reptile with steamlined body and long snout

ichthyostega (IK-the-eh-ste-ga)   an extinct primitive Paleozoic fishlike amphibian

ignimbrite (IG-nem-brite)   a hard rock composed of consolidated pyroclastic material

ilmenite (IL-meh-ite)   a black iron-titanium ore

index fossil   a representative fossil that identifies the rock stratum in which it is found

interglacial   a warming period between glacial periods

intrusive   a granitic body that invades the Earth's crust

invertebrate   an animal with an external skeleton such as a shellfish or insect

iridium   a rare isotope of platinum, relatively abundant on meteorites

island arc   volcanoes landward of a subduction zone parallel to a trench and above the melting zone of a subducting plate

kaolinite (KAY-oh-lin-ite)   a hydrous aluminum silicate clay from the decay of feldspar

karst   a terrain that comprises numerous sinkholes in limestone

kimberlite   an agglomerate biotite-periodotite occurring in volcanic pipes containing diamonds

lacustrine (leh-KES-trene)   inhabiting or produced in lakes

lapilli   gravellike pyroclastic deposits

Laurasia (LURE-ay-zha)   a northern supercontinent of Paleozoic time, consisting of North America, Europe, and Asia

Laurentia (LURE-in-tia)   an ancient North American continent

lava   molten magma that flows out onto the surface

limestone   a sedimentary rock composed of calcium carbonate that is secreted from seawater by invertebrates and whose skeletons composed the bulk of deposits

lithospheric   a segment of the lithosphere, the upper layer plate of the mantle, involved in the interaction of other plates in tectonic activity

loess   a thick deposit of airborne dust

lungfish   a bony fish that breathes on land and in water

lycopod (LIE-keh-pod)   the first ancient trees of Paleozoic forests; today comprising club mosses and liverworts

mafic   a dark iron-magnesium silicate mineral

magma   a molten rock material generated within the Earth and the constituent of igneous rocks

magnetic field reversal   a reversal of the north-south polarity of the Earth's magnetic poles

magnitite   a black iron-oxide ore

mantle   the part of a planet below the crust and above the core, composed of dense rocks that might be in convective flow

marsupial (mar-SUE-pee-al)   a primitive mammal that weans underdeveloped infants in a belly pouch

Mesozoic (MEH-zeh-ZOE-ik)   literally the time of middle life, referring to a period between 250 and 65 million years ago

metamorphism   recrystallization of previous igneous, metamorphic, and sedimentary rocks under extreme temperatures and pressures without melting

metazoan   primitive multicellular animal with cells differentiated for specific functions

meteoritics   the study of meteors and meteorites

mica   colored to clear silicate with perfect cleavage able to split into thin sheets

micrite   a microcrystalline matrix of limestone

microcline   an orthoclase feldspar found in pegmatites and metamorphic rocks

microcrystalline   having crystals invisible to the unaided eye

microfossil   a fossil that must be studied with a microscope; used for dating drill cuttings

midocean ridge   a submarine ridge along a divergent plate boundary where a new ocean floor is created by the upwelling of mantle material

mold   an impression of a fossil shell or other organic structure made in encasing material

mollusk (MAH-lusk)   a large group of invertebrates, including snails, clams, squids, and extinct ammonites, characterized by an internal and an external shell surrounding the body

monotreme   egg-laying mammals including platypus and echidna

moraine   a ridge of erosional debris deposited by the melting margin of a glacier

nautiloid (NOT-eh-loid)   shell-bearing cephalopods abundant in the Paleozoic of which only the nautilus survived

Neogene   the Miocene and Pliocene epochs of the Cenozoic

nepheline (NE-fe-lene)   a sodium, potassium, and aluminum silicate common in igneous rocks

nuée ardente (NU-ee ARE-dent)   an avalanche of glowing clouds of ash and pyroclastics

olivine (AH-leh-vene)   a green iron-magnesium silicate common in intrusive and volcanic rocks

oolite (OH-eh-lite)   small rounded grains in limestone

ophiolite (OH-fi-ah-lite)   oceanic crust thrust upon continents by plate tectonics

orogeny (oh-RAH-ja-nee)   an episode of mountain building by tectonic activity

orthoclase (OR-the-clase)   a white, gray, or pink potassium feldspar

pahoehoe (pah-HOE-ay-hoe-ay)   a Hawaiian term for ropy basalt lava

Paleogene   the Paleocene, Eocene, and Oligocene epochs of the Cenozoic

paleomagnetism   the study of the Earth's magnetic field, including the position and polarity of the poles in the past

paleontology (PAY-lee-ON-tah-logy)   the study of ancient life-forms, based on the fossil record of plants and animals

Paleozoic (PAY-lee-eh-ZOE-ic)   the period of ancient life, between 570 and 250 million years ago

Pangaea (PAN-gee-ah)   a Paleozoic supercontinent that included all the lands of the Earth

Panthalassa (PAN-the-lass-ah)   the global ocean that surrounded Pangaea

pegmatite   an igneous rock body with extremely large crystals, often enriched with ores

period   a division of geologic time longer than an epoch and included in an era

permafrost   permanently frozen ground in the Arctic regions

phyla   groups of organisms that share similar body forms

placoderm   an extinct class of chordates, fish with armorlike plates and articulated jaws

plagioclase (PLAY-jee-eh-klase)   a calcium or sodium feldspar

plate tectonics   the theory that accounts for the major features of the Earth's surface in terms of the interaction of lithospheric plates

precipitation   the deposition of minerals from seawater

prokaryote (pro-KAR-ee-ote)   a primitive organism that lacks a nucleus

protistid (PRO-tist-id)   a unicellular organism, including bacteria, protozoans, algae, and fungi

pseudofossil   a fossillike body such as a concretion

pterosaur (TER-eh-sore)   extinct Mesozoic flying reptiles with batlike wings

pyroclastic   fragments produced by volcanic eruptions

pyroxene (pie-ROCK-seen)   common igneous rock–forming silicate minerals with short, dark prismatic crystals

quartzite   an extremely hard metamorphosed sandstone or a sandstone whose grains are cemented with silica

radiolarian   a microorganism with shells made of silica constituting a large component of siliceous sediments

radiometric dating   the age determination of an object by chemical and radiometric analysis of stable versus unstable radioactive elements

redbed   a sedimentary rock cemented with iron oxide

reef   the biologic community that lives at the edge of an island or continent; the shells of dead organisms form a limestone deposit

regression   a fall in sea level, exposing continental shelves to erosion

reptile   an air-breathing, cold-blooded animal that lays eggs and is usually covered with scales

rhyolite (RYE-eh-lite)   a potassium feldspar-rich volcanic rock equivalent to granite

Rodinia   a Precambrian supercontinent whose breakup sparked the Cambrian explosion of species

sandstone   a sedimentary rock consisting of sand grains cemented together

schist (shist)   a finely layered metamorphic crystalline rock easily split along parallel bands

seafloor spreading   a theory that the ocean floor is created by the separation of lithospheric plates along midocean ridges, with new oceanic crust formed from mantle material that rises from the mantle to fill the rift

shale   a fine-grained fissile sedimentary rock of consolidated mud or clay

shield   areas of the exposed Precambrian nucleus of a continent

sial   a rock rich in silica and aluminum in continental crust

sill   a tabular igneous intrusive with boundaries parallel to the planar structure of the surrounding rock

sima   a rock rich in silica and magnesium in oceanic crust

species   groups of organisms that share similar characteristics and are able to breed among themselves

sphalerite (SFA-leh-rite)   a red to brown iron-zinc sulfide ore

spherules   small spherical glassy grains found on certain types of meteorites, in lunar soils, and at large meteorite impact sites

spodumene (SPA-jeh-mene)   a lithium-aluminum silicate ore of the pyroxene group

stock   an intrusive body of a deep-seated igneous rock, usually lacking conformity and resembling a batholith with smaller dimensions

strata   layered rock formations; also called beds

stromatolite (STRO-mat-eh-lite)   a calcareous structure built by successive layers of bacteria or algae that has existed for the past 3.5 billion years

subduction zone   a region where an oceanic plate dives below a continental plate into the mantle; ocean trenches are the surface expression of a subduction zone

superconductivity   the loss of electrical resistance at low temperatures

tectonic activity   the formation of the Earth's crust by large-scale movements throughout geologic time

tektites   small glassy minerals created from the melting of surface rocks by the impact of a large meteorite

tephra (TE-fra)   solid material ejected into the air by a volcanic eruption

terrestrial   all phenomena pertaining to the Earth

Tethys Sea (TEH-this)   the hypothetical midlatitude region of the oceans separating the northern and southern continents of Laurasia and Gondwana several hundred million years ago

tetrapod   a four-footed vertebrate

thecodont (THEE-keh-daunt)   an ancient primitive reptile that gave rise to dinosaurs, crocodiles, and birds

therapsid (the-RAP-sid)   ancient reptile that was the ancestor of the mammals

therian   animals that give live births such as mammals

thermophilic   relating to primitive organisms that live in hot water environments

till   nonstratified material deposited directly by glacial ice as it recedes and consolidated into tillite

transgression   a rise in sea level that causes flooding of the shallow edges of continental margins

trilobite (TRY-leh-bite)   an extinct marine arthropod, characterized by a body divided into three lobes, each bearing a pair of jointed appendages, and a chitinous exoskeleton

tuff   a consolidated fine-grained pyroclastic volcanic rock

tundra   permanently frozen ground at high latitudes

type section   a sequence of strata that was originally described as constituting a stratigraphic unit and serves as a standard of comparison for identifying similar widely separated units

ultraviolet   the invisible light with a wavelength shorter than that of visible light and longer than that of X rays

uniformitarianism   a theory that the slow processes that shape the Earth's surface have acted essentially unchanged throughout geologic time

uraninite   a black uranium oxide ore

varves   thinly laminated lake bed sediments deposited by glacial meltwater

vein   a tabular or sheetlike body of one or more minerals deposited in openings of fissures, joints, or faults

vertebrates   animals with an internal skeleton, including fish, amphibians, reptiles, and mammals

zeolite   a group of hydrous silicates similar in composition to feldspars often found in cavities in lava

zircon   a silicate of zirconium whose crystals are sometimes cut as gems and used to date the oldest rocks

BIBLIOGRAPHY

THE EARTH'S HISTORY

Allegre, Claude J. and Stephen H. Schneider. "The Evolution of the Earth," *Scientific American* 271 (October 1994): 66–75.

Boucot, A.J. and Jane Gray. "A Paleozoic Pangaea," *Science* 222 (November 11, 1983): 571–580.

Dalziel, Ian W.D. "Earth before Pangaea," *Scientific American* 272 (January 1995): 58–63.

Dupré, Roslyn M. "Earth's Early Evolution," *Earth* 5 (February 1996): 14.

Eldredge, Niles. *Life Pulse: Episodes from the Story of the Fossil Record.* New York: Facts On File, 1987.

Jones, Richard C. and Anthony N. Stranges. "Unraveling Origins, the Archean," *Earth Science* 42 (Winter 1989): 20–22.

Knoll, Andrew H. "End of the Proterozoic Eon," *Scientific American* 265 (October 1991): 64–73.

Matthewes, Samual W. "Ice on the World," *National Geographic* 171 (January 1987): 84–103.

Moores, Eldridge. "The Story of Earth," *Earth* 5 (December 1996): 30–33.

Simpson, Sarah. "Life's First Scalding Steps," *Science News* 155 (January 9, 1999): 24–26.

Stokes, W. Lee. *Essentials of Earth History.* Englewood Cliffs, N.J.: Prentice-Hall, 1982.

Taylor, S. Ross and Scott M. McLennan. "The Evolution of Continental Crust," *Scientific American* 274 (January 1996): 76–81.

Vogel, Shawna. "Living Planet," *Earth* 5 (April 1996): 27–35.

Waters, Tom. "Greetings from Pangaea," *Discover* 13 (February 1992): 38–43.

Wess, Peter. "Land before Time," *Earth* 8 (February 1998): 29–33.

CLUES TO THE PAST

Badash, Lawrence. "The Age-of-the-Earth Debate," *Scientific American* 261 (August 1989): 90–96.

Beardsley, Tim. "Punctuated Equilibrium," *Scientific American* 262 (March 1990): 36–38.

Benton, Michael J. "Late Triassic Extinctions and the Origin of the Dinosaurs," *Science* 260 (May 7, 1993): 769–770.

Boling, Rick, "The Mystery of Selective Extinctions," *Earth* 5 (October 1996): 12.

Eldredge, Niles. "What Drives Evolution?" *Earth* 5 (December 1996): 34–37.

Erwin, Douglas H. "The Mother of Mass Extinctions," *Scientific American* 275 (July 1996): 72–78.

Gould, Stephen Jay. "The Evolution of Life on the Earth." *Scientific American* 271 (October 1994): 85–91.

Herbert, Sandra. "Darwin as a Geologist," *Scientific American* 254 (May 1986): 116–123.

Raup, David M. "Biological Extinction in Earth History," *Science* 231 (March 28, 1986): 1528–1533.

Stebbins G. Ledyard and Francisco J. Ayala. "The Evolution of Darwinism," *Scientific American* 253 (July 1985): 72–82.

Waldrop, Mitchell M. "Spontaneous Order, Evolution and Life." *Science* 247 (March 30, 1990): 1543–1545.

York, Derek. "The Earliest History of the Earth," *Scientific American* 268 (January 1993): 90–96.

ROCK TYPES

Berner, Robert A. and Antonio, C. Lasaga. "Modeling the Geochemical Carbon Cycle," *Scientific American* 260 (March 1989): 74–81.

Coffin, Millard F. and Olav Eldholm. "Large Igneous Provinces," *Scientific American* 269 (October 1993): 42–49.

Decker, Robert and Barbara Decker. *Volcanoes.* San Francisco: Freeman, 1981.

Galer, Stephen J.G. "Oldest Rocks in Europe," *Nature* 370 (August 18, 1994): 505–506.

Gore, Rick. "Living on Fire," *National Geographic* 173 (May 1998): 4–37.

Green, D. H., S. M. Eggins, and G. Yaxley. "The Other Carbon Cycle," *Nature* 365 (September 16, 1993): 210–211.

Howell, David G. "Terranes," *Scientific American* 253 (November 1985): 116–125.

Mack, Walter N. and Elizabeth A. Leistikow. "Sands of the World," *Scientific American* 275 (August 1996): 62–67.

Pendick, Daniel. "Ashes, Ashes, All Fall Down," *Earth* 6 (February 1997): 32–33.

Pennisi, Elizabeth. "Dancing Dust," *Science News* 142 (October 3, 1992): 218–220.

Peterson, Ivars. "Digging into Sand," *Science News* 136 (July 15, 1989): 40–42.

Preiser, Rachel. "The Dolomite Problem," *Discover* 16 (February 1996): 32.

Ritchie, David. *The Ring of Fire.* New York: Antheneum, 1981.

Vink, Gregory E., et al. "The Earth's Hot Spots," *Scientific American* 252 (April 1985): 50–57.

FOSSIL FORMATION

Burgin, Toni et. al. "The Fossils of Monte San Giorgio," *Scientific American* 260 (June 1989): 74–81.

Grimaldi, David A. "Captured in Amber," *Scientific American* 274 (April 1996): 84–91.

Jeffery, David. "Fossils: Animals of Life Written in Rock," *National Geographic* 168 (August 1985): 182–191.

Monastersky, Richard. "Is Bigger Better? The Fossils Speak Up," *Science News* 151 (February 1, 1997): 72.

Mossman, David J. and William A.S. Sarjeant. "The Footprints of Extinct Animals," *Scientific American* 248 (January 1983): 75–85.

Newton, Cathryn R. "Significance of Tethyan Fossils in the American Cordillera," *Science* 242 (October 21, 1988): 385–390.

Novacek, Michael J., et al. "Fossils of the Flaming Cliffs," *Scientific American* 271 (December 1994): 60–69.

Pinna, Giovanni. *The Illustrated Encyclopedia of Fossils.* New York: Facts On File, 1990.

Ridley, Mark. "Evolution and Gaps in the Fossil Record," *Nature* 286 (July 31, 1980): 444–445.

Schreeve, James. "Are Algae—Not Coral—Reefs' Master Builders?" *Science* 271 (February 2, 1996): 597–598.

Watson, Andrew. "Will Fossils from Down Under Upend Mammal Evolution?" *Science* 278 (November 21, 1997): 1401.

Wright, Karen. "What the Dinosaurs Left Us," *Discover* 17 (June 1996): 59–65.

MARINE FOSSILS

Duve, Christian. "The Birth of Complex Cells," *Scientific American* 274 (April 1996): 50–57.

Forey, Peter and Philippe Janvier. "Agnathans and the Origin of Jawed Vertebrates," *Nature* 361 (January 14, 1993): 129–133.

Irion, Robert. "Parsing the Trilobites' Rise and Fall," *Science* 280 (June 19, 1998): 1837.

Levinton, Jeffrey S. "The Big Bang of Animal Evolution," *Scientific American* 267 (November 1992): 84–91.

McMenamin, Mark A.S. "The Emergence of Animals," *Scientific American* 256 (April 1987): 94–102.

Monastersky, Richard. "The Rise of Life on Earth," *National Geographic* 194 (March 1998): 54–81.

Morris, S. Conway. "Burgess Shale Faunas and the Cambrian Explosion," *Science* 246 (October 20, 1989): 339–345.

Radetsky, Peter. "Life's Crucible," *Earth* 7 (February 1998): 34–41.

Richardson, Joice R. "Brachiopods," *Scientific American* 255 (September 1985): 100–106.

Svitil, Kathy A. "It's Alive, and It's Graptolite," *Discover* 14 (July 1993): 18–19.

Ward, Peter. "The Extinction of the Ammonites," *Scientific American* 249 (October 1983): 136–147.

Wright, Karen. "When Life Was Odd," *Discover* 18 (March 1997): 52–61.

TERRESTRIAL FOSSILS

Culotta, Elizabeth. "Ninety Ways to Be a Mammal," *Science* 266 (November 18, 1994): 1161.

Fischman, Josh. "Dino Hunter," *Discover* 20 (May 1999): 72–78.

Horner, John R. "The Nesting Behavior of Dinosaurs," *Scientific American* 250 (April 1984): 130–137.

Langston, Wann, Jr. "Pterosaurs," *Scientific American* 244 (February 1981): 122–136.

Morell, Virginia. "Announcing the Birth of a Heresy," *Discover* 8 (March 1987): 26–51.

Padian, Kevin and Luis M. Chiappe. "The Origin of Birds and Their Flight," *Scientific American* 218 (February 1998): 38–47.

Pendick, Daniel. "The Mammal Mother Lode," *Earth* 4 (April 1995): 20–23.

Robbins, Jim. "The Real Jurassic Park," *Discover* 12 (March 1991): 52–59.

Schmidt, Karen. "Rise of the Mammals," *Earth* 5 (October 1996): 20–21, 68–69.

Schueller, Gretel. "Mammal 'Missing Link' Found," *Earth* 7 (April 1998): 9.

Storch, Gerhard. "The Mammals of Island Europe," *Scientific American* 266 (February 1992): 64–69.

Vickers-Rich, Patricia and Thomas Hewitt Rich. "Polar Dinosaurs of Australia," *Scientific American* 269 (July 1993): 49–55.

Wellnhofer, Peter. "Archaeopteryx," *Scientific American* 262 (May 1990): 70–77.

Zimmer, Carl. "Coming onto the Land," *Discover* 16 (June 1995): 120–127.

CRYSTALS AND MINERALS

Allen, Joseph B. "New Mineral: Out of the Blue," *Earth* 5 (August 1996): 18.

Barnes, H. L. and A. W. Rose. "Origins of Hydrothermal Ores," *Science* 279 (March 27, 1998): 2064–2065.

Brinhall, George. "The Genesis of Ores," *Scientific American* 264 (May 1991): 84–91.

Gass, Ian G. "Ophiolites," *Scientific American* 274 (August 1982): 122–131.

Hazen, Robert M. and Larry W. Finger. "Crystals at High Pressure," *Scientific American* 252 (May 1985): 100–117.

Nelson, David R. "Quasicrystals," *Scientific American* 255 (August 1996): 43–51.

Peterson, Ivars. "Tilings for Picture-Perfect Quasicrystals," *Science News* 137 (January 13, 1990): 22.

Rona, Peter A. "Mineral Deposits from Sea-Floor Hot Springs," *Scientific American*. 254 (January 1986): 84–92.

Scovil, Jeff. "Minerals in Disguise," *Earth* 6 (October 1997): 11.

Weisburd, Stefi. "Meteor Linked to Rich Ores at Sudbury," *Science News* 128 (October 26, 1985): 263.

Wu, Corinna, "Gazing into Crystal Balls," *Science News* 151 (April 12, 1997): 224–225.

———. "How Zeolites Hold Tight to Metal Ions," *Science News* 151 (May 24, 1997): 319.

———. "Impurities Give Crystals That Special Glow," *Science News* 151 (May 17, 1997): 303.

Yardley, Bruce W.D. "Rocks of Rare Refinements," *Nature* 369 (May 5, 1994): 17.

GEMS AND PRECIOUS METALS

Cox, Keith G. "Kimberlite Pipes," *Scientific American* 238 (April 1978): 120–132.

Edward Cayce Foundation. *Gems and Stones.* Virginia Beach, Va.: A.R.E. Press, 1987.

Green, Timothy. "All That Glitters," *Modern Maturity* 31 (August/September 1988): 54–60.

Hall, Cally, *Gem Stones.* New York: Dorling Kindersley, 1994.

Kaiser, Jocelyn. "Insulator's Baby Steps to Superconductivity." *Science* 282 (December 11, 1998): 1966–1967.

Kerr, Richard A. "Bits of the Lower Mantle Found in Brazilian Diamonds," *Science* 261 (September 10, 1993): 1391.

Monastersky, Richard. "Microscopic Diamonds Crack Geologic Mold," *Science News* 148 (July 8, 1995): 22.

Muecke, Gunter K. and Peter Moller. "The Not-So-Rare Earths," *Scientific American* 258 (January 1988): 72–77.

Pendick, Daniel. "Amber-Trapped Creatures Show Timeless Form," *Science News* 143 (January 16, 1993): 39.

Peterson, Ivars. "Growing a Fibrous Superconductor," *Science News* 133 (June 25, 1988): 406.

Schueller, Gretel. "Plumes of Gold," *Earth* 6 (December 1997): 13–14.

Schumann, Walter. *Gemstones of the World.* New York: Sterling, 1997.

Witzke, Brian J. "Geodes from Iowa," *Earth Science* 41 (Summer 1988): 19.

THE RARE AND UNUSUAL

Cowen, Ron. "After the Fall," *Science News* 148 (October 14, 1995): 248–249.

Fischman, Joshua. "Flipping the Field," *Discover* 11 (May 1990): 28–29.

Fryer, Patricia. "Mud Volcanoes of the Marianas," *Scientific American* 266 (February 1992): 46–52.

Hecht, Jeff. "Death Valley Rocks Skate on Thin Ice," *New Scientist* 248 (September 20, 1995): 19.

Krantz, William B., et al. "Pattern Ground," *Scientific American* 259 (December 1988): 68–76.

Monastersky, Richard. "The Light at the Bottom of the Ocean," *Science News* 150 (September 7, 1996): 156–157.

Nori, Franco, et al. "Booming Sand," *Scientific American* 277 (September 1997): 84–89.

Perth, Nigel Henbest "Meteorite Bonanza in Australian Desert," *New Scientist* 129 (April 20, 1991): 20.

Schueller, Gretel. "One-note Eruptions," *Earth* 8 (February 1998): 14.

Sharpton, Virgil L. "Glasses Sharpen Impact Views," *Geotimes* 33 (June 1988): 10–11.

Weisburd, Stefi. "Largest Melt from Lightning Strike," *Science News* 130 (October 11, 1986): 108–110.

———. "The Microbes That Loved the Sun," *Science News* 129 (February 15, 1986): 108–110.

————. "Self-Reversing Minerals Make a Comeback," *Science News* 127 (April 13, 1985): 234–236.

WHERE FOSSILS AND MINERALS ARE FOUND

Averett, Walter R. "Fertile Fossil Field," *Earth Science* 41 (Spring 1988): 16–18.

Birnbraum, Stephen. "The Grand Canyon," *Good Housekeeping* (November 1990): 150–152.

Colbert, Edward H. *A Fossil Hunter's Notebook.* New York: E.P. Dutton, 1980.

DiChristina, Mariette. "The Dinosaur Hunter," *Popular Science* (September 1996): 41–46.

Fisher, Louis J. "Finding Fossils," *Earth Science* 41 (Summer 1988): 20–22.

Friedman, Gerald M. "Slides and Slumps," *Earth Science* 41 (Fall 1988): 21–23.

Goetz, Alexander F.H. and Lawrence C. Rowan. "Geologic Remote Sensing," *Science* 211 (February 20, 1981): 781–790.

Hannibal, Joseph T. "Quarries Yield Rare Paleozoic Fossils," *Geotimes* 33 (July 1988): 10–13.

Lockwood, C. C. "Wonder Holes," *International Wildlife* 20 (January/February 1990): 47–49.

MacFall, Russell P. *Rock Hunter's Guide.* New York: Thomas Y. Crowell, 1980.

Maslowski, Andy. "Eyes on the Earth," *Astronomy* 14, (August 1986): 9–10.

Miller, Martin. "Missing Time," *Earth* 4 (October 1995): 58–60.

Monastersky, Richard. "What's New in the Ol' Grand?" *Science News* 132 (December 19, 26, 1987): 392–395.

Wood, Dennis. "The Power of Maps," *Scientific American* 268 (May 1993): 89–93.

INDEX

Boldface page numbers indicate extensive treatment of a topic. *Italic* page numbers indicate illustrations or captions. Page numbers followed by *m* indicate maps; *t* indicate tables; *g* indicate glossary.

Yellowstone River, Wyoming
173
Yosemite, California *223*
ytterbium 193
yttrium 192, 193

Z

zeolites 159, 224, 252g
zinc 162, 165, 166, 209
zircon *177*, 224, 252g

zirconium 178
zooanthellae algae 107